Letts

Revise
AS

ab^2

Edexcel
Mathematics

Contents

Chapter 1 Core 1 Pure Mathematics

Chapter 2 Core 2 Pure Mathematics

Specification information

Edexcel Mathematics

MODULE	SPECIFICATION TOPIC	CHAPTER REFERENCE	STUDIED IN CLASS	REVISED	PRACTICE QUESTIONS
Core 1	Algebra and functions	1.1			
	Coordinate geometry in the (x, y) plane	1.2			
	Sequences and series	1.3			
	Differentiation	1.4			
	Integration	1.5			
Core 2	Algebra and functions	2.1			
	Coordinate geometry in the (x, y) plane	2.2			
	Sequences and series	2.3			
	Trigonometry	2.4			
	Exponentials and logarithms	2.5			
	Differentiation	2.6			
	Integration	2.7			
Mechanics 1	Vectors in mechanics	3.1			
	Kinematics in a straight line	3.2			
	Statics of a particle	3.3			
	Moments	3.3			
	Dynamics in a straight line or plane	3.4			
Statistics 1	Representation and summary of data	4.1			
	Probability	4.2			
	Correlation and regression	4.5			
	Discrete random variables	4.3			
	The normal distribution	4.4			
Decision 1	Algorithms	5.1			
	Algorithms on graphs	5.2			
	The route inspection problem	5.2			
	Critical path analysis	5.3			
	Linear programming	5.4			
	Matchings	5.5			

Examination analysis

The *assessment is by written papers. All questions are compulsory.*
Core 1 + Core 2 + one of Mechanics 1, Statistics 1 and Decision 1.

Core 1	AS	No calculator	1 hr 30 min exam	33.3%
Core 2	AS	Scientific / graphic calculator	1 hr 30 min exam	33.3%
Mechanics 1	AS	Scientific / graphic calculator	1 hr 30 min exam	33.3%
Statistics 1	AS	Scientific / graphic calculator	1 hr 30 min exam	33.3%
Decision 1	AS	Scientific / graphic calculator	1 hr 30 min exam	33.3%

Your AS/A2 Level Mathematics course

AS and A2

The Edexcel Mathematics A Level course is made up of six modules. Students first study the AS (Advanced Subsidiary) course, which has three modules. Some will then go on to study the second part of the A Level course, a further three modules, called A2. Advanced Subsidiary is assessed at the standard expected halfway through an A Level course: i.e., between GCSE and Advanced GCE. This means that AS and A2 courses are designed so that difficulty steadily increases:

- AS Mathematics builds from GCSE Mathematics
- A2 Mathematics builds from AS Mathematics.

How will you be tested?

Assessment units

For AS Mathematics, you will be tested by three assessment units. For the full A Level in Mathematics, you will take a further three units. AS Mathematics forms 50% of the assessment weighting for the full A Level.

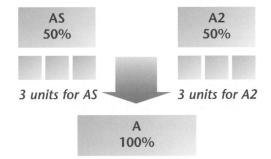

3 units for AS   3 units for A2

Each unit can normally be taken in either January or June. Alternatively, you can study the whole course before taking any of the unit tests. There is a lot of flexibility about when exams can be taken and the diagram below shows just some of the ways that the assessment units may be taken for AS and A Level Mathematics.

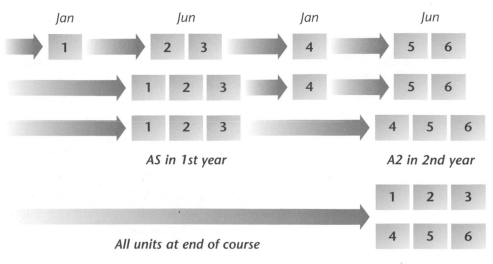

If you are disappointed with a module result, you can resit the module. There is no restriction on the number of times a module may be attempted. The best available result for each module will count towards the final grade.

Synoptic assessment

The GCE Advanced Subsidiary and Advanced Level Qualification specific Criteria state that A Level specifications must include synoptic assessment (representing at least 20% of the total A Level marks).

Synoptic assessment in mathematics addresses candidates' understanding of the connections between different elements of the subject. It involves the explicit drawing together of knowledge, understanding and skills learned in different parts of the A Level course, through using and applying methods developed at earlier stages of the course to solving problems. Making and understanding connections in this way is intrinsic to learning mathematics.

Key skills

It is important that you develop your key skills throughout your AS and A2 courses. These are important skills that you need whatever you do beyond AS and A Levels. To gain the key skills qualification, which is equivalent to an AS Level, you will need to collect evidence together in a 'portfolio' to show that you have attained Level 3 in Communication, Application of number and Information technology. You will also need to take a formal test in each key skill. You will have many opportunities during AS Mathematics to develop your key skills.

It is a worthwhile qualification, as it demonstrates your ability to put your ideas across to other people, collect data and use up-to-date technology in your work.

What skills will I need?

For AS Mathematics, you will be tested by assessment objectives: these are the skills and abilities that you should have acquired by studying the course. The assessment objectives for AS Mathematics are shown below.

Candidates should be able to:

- recall, select and use their knowledge of mathematical facts, concepts and techniques in a variety of contexts
- construct rigorous mathematical arguments and proofs through use of precise statements, logical deduction and inference and by the manipulation of mathematical expressions, including the construction of extended arguments for handling substantial problems presented in unstructured form
- recall, select and use their knowledge of standard mathematical models to represent situations in the real world; recognise and understand given representations involving standard models; present and interpret results from such models in terms of the original situation, including discussion of the assumptions made and refinement of such models
- comprehend translations of common realistic contexts into Mathematics; use the results of calculations to make predictions, or comment on the context; and, where appropriate, read critically and comprehend longer mathematical arguments or examples of applications

- use contemporary calculator technology and other permitted resources (such as formulae booklets or statistical tables*) accurately and efficiently; understand when not to use such technology, and its limitations; give answers to appropriate accuracy.

  * A copy of the formulae booklet and statistical tables can be found on the A Level mathematics section of the Edexcel website, www.edexcel.com

Progression and prior learning

Mathematics is, inherently, a sequential subject. There is a progression of material through all levels at which the subject is studied. The criteria therefore build on the knowledge, understanding and skills established at GCSE.

Thus, candidates embarking on AS/A2 Level study in Mathematics subjects are expected to have covered all the material in the Higher Tier. This material is regarded as assumed background knowledge. However, it may be assessed within questions focused on other material.

Exam technique

What are the examiners looking for?

Examiners use certain words in their instructions to let you know what they are expecting in your answer. Make sure that you know what they mean so that you can give the right response.

Write down, state

You can write your answer without having to show how it was obtained. There is nothing to prevent you doing some working if it helps you, but if you are doing a lot then you might have missed the point.

Calculate, find, determine, show, solve

Make sure that you show enough working to justify the final answer or conclusion. Marks will be available for showing a correct method.

Deduce, hence

This means that you are expected to use the given result to establish something new. You must show all of the steps in your working.

Draw

This is used to tell you to plot an accurate graph using graph paper. Take note of any instructions about the scale that must be used. You may need to read values from your graph.

Sketch

If the instruction is to sketch a graph then you don't need to plot the points but you will be expected to show its general shape and its relationship with the axes. Indicate the positions of any turning points and take particular care with any asymptotes.

Find the exact value

This instruction is usually given when the final answer involves an irrational value such as a logarithm, π or a surd. You will need to demonstrate that you can manipulate these quantities so don't just key everything into your calculator or you will lose marks.

If a question requires the final answer to be given to a specific level of accuracy then make sure that you do this or you might needlessly lose marks.

Some dos and don'ts

Dos

Do read the question

- Make sure that you are clear about what you are expected to do. Look for some structure in the question that may help you take the right approach.
- Read the question *again* after you have answered it as a quick check that your answer is in the expected form.

Do use diagrams

- In some questions, particularly in mechanics, a clearly labelled diagram is essential. Use a diagram whenever it may help you understand or represent the problem that you are trying to solve.

Do take care with notation

- Write clearly and use the notation accurately. Use brackets when they are required.
- Even if your final answer is wrong, you may earn some marks for a correct expression in your working.

Do avoid silly answers

- Check that your final answer is sensible within the context of the question.

Do make good use of time

- Choose the order in which you answer the questions carefully. Do the ones you find easiest first.
- Set yourself a time limit for a question depending on the number of marks available.
- Be prepared to leave a difficult part of a question and return to it later if there is time.
- Towards the end of the exam make sure that you pick up all of the easy marks in any questions that you haven't got time to answer fully.

Don'ts

Don't work with rounded values

- There may be several stages in a solution that produce numerical values. Rounding errors from earlier stages may distort your final answer. One way to avoid this is to make use of your calculator memories to store values that you will need again.

Don't cross out work that may be partly correct

- It's tempting to cross out something that hasn't worked out as it should. Avoid this unless you have time to replace it with something better.

Don't write out the question

- This wastes time. The marks are for your solution!

What grade do you want?

Everyone should be able to improve their grades but you will only manage this with a lot of hard work and determination. The details given below describe a level of performance typical of candidates achieving grades A, C or E. You should find it useful to read and compare the expectations for the different levels and to give some thought to the areas where you need to improve most.

Grade A candidates

- Recall or recognise almost all the mathematical facts, concepts and techniques that are needed, and select appropriate ones to use in a variety of contexts.
- Manipulate mathematical expressions and use graphs, sketches and diagrams, all with high accuracy and skill.
- Use mathematical language correctly and proceed logically and rigorously through extended arguments or proofs.
- When confronted with unstructured problems they can often devise and implement an effective solution strategy.
- If errors are made in their calculations or logic, these are sometimes noticed and corrected.
- Recall or recognise almost all the standard models that are needed, and select appropriate ones to represent a wide variety of situations in the real world.
- Correctly refer results from calculations using the model to the original situation; they give sensible interpretations of their results in the context of the original realistic situation.
- Make intelligent comments on the modelling assumptions and possible refinements to the model.
- Comprehend or understand the meaning of almost all translations into mathematics of common realistic contexts.
- Correctly refer the results of calculations back to given context and usually make sensible comments or predictions.
- Can distil the essential mathematical information from extended pieces of prose having mathematical content.
- Comment meaningfully on the mathematical information.
- Make appropriate and efficient use of contemporary calculator technology and other permitted resources, and are aware of any limitations to their use.
- Present results to an appropriate degree of accuracy.

Grade C candidates

- Recall or recognise most of the mathematical facts, concepts and techniques that are needed, and usually select appropriate ones to use in a variety of contexts.
- Manipulate mathematical expressions and use graphs, sketches and diagrams, all with a reasonable level of accuracy and skill.
- Use mathematical language with some skill and sometimes proceed logically through extended arguments or proofs.
- When confronted with unstructured problems they sometimes devise and implement an effective and efficient solution strategy.
- Occasionally notice and correct errors in their calculations.
- Recall or recognise most of the standard models that are needed and usually select appropriate ones to represent a variety of situations in the real world.

- Often correctly refer results from calculations using the model to the original situation; they sometimes give sensible interpretations of their results in context of the original realistic situation.
- Sometimes make intelligent comments on the modelling assumptions and possible refinements to the model.
- Comprehend or understand the meaning of most translations into mathematics of common realistic contexts.
- Often correctly refer the results of calculations back to the given context and sometimes make sensible comments or predictions.
- Distil much of the essential mathematical information from extended pieces of prose having mathematical content.
- Give some useful comments on this mathematical information.
- Usually make appropriate and effective use of contemporary calculator technology and other permitted resources, and are sometimes aware of any limitations to their use.
- Usually present results to an appropriate degree of accuracy.

Grade E candidates

- Recall or recognise some of the mathematical facts, concepts and techniques that are needed, and sometimes select appropriate ones to use in some contexts.
- Manipulate mathematical expressions and use graphs, sketches and diagrams, all with some accuracy and skill.
- Sometimes use mathematical language correctly and occasionally proceed logically through extended arguments or proofs.
- Recall or recognise some of the standard models that are needed and sometimes select appropriate ones to represent a variety of situations in the real world.
- Sometimes correctly refer results from calculations using the model to the original situation; they try to interpret their results in the context of the original realistic situation.
- Sometimes comprehend or understand the meaning of translations into mathematics of common realistic contexts.
- Sometimes correctly refer the results of calculations back to the given context and attempt to give comments or predictions.
- Distil some of the essential mathematical information from extended pieces of prose having mathematical content; they attempt to comment on this mathematical information.
- Candidates often make appropriate and efficient use of contemporary calculator technology and other permitted resources.
- Often present results to an appropriate degree of accuracy.

The table below shows how your uniform standardised mark is translated.

uniform mark (max. 100)	80	70	60	50	40
grade	A	B	C	D	E

The A* will be awarded to students who have achieved a grade A overall (480 UMS or more) and 180 UMS or more on the total of their Core 3 and Core 4 units. It is awarded for A Level qualification only and not for the AS qualification or individual units.

Four steps to successful revision

Step 1: Understand

- Study the topic to be learned slowly. Make sure you understand the logic or important concepts.
- Mark up the text if necessary – underline, highlight and make notes.
- Re-read each paragraph slowly.

GO TO STEP 2

Step 2: Summarise

- Now make your own revision note summary:
  What is the main idea, theme or concept to be learned?
  What are the main points? How does the logic develop?
  Ask questions: Why? How? What next?
- Use bullet points, mind maps, patterned notes.
- Link ideas with mnemonics, mind maps, crazy stories.
- Note the title and date of the revision notes
  (e.g. Mathematics: Trigonometry, 3rd March).
- Organise your notes carefully and keep them in a file.

This is now in **short-term memory**. You will forget 80% of it if you do not go to Step 3.
GO TO STEP 3, but first take a 10 minute break.

Step 3: Memorise

- Take 25 minute learning 'bites' with 5 minute breaks.
- After each 5 minute break test yourself:
  Cover the original revision note summary
  Write down the main points
  Speak out loud (record yourself)
  Tell someone else
  Repeat many times.

The material is well on its way to **long-term memory**.
You will forget 40% if you do not do step 4. *GO TO STEP 4*

Step 4: Track/Review

- Create a Revision Diary (one A4 page per day).
- Make a revision plan for the topic, e.g. 1 day later, 1 week later, 1 month later.
- Record your revision in your Revision Diary, e.g.
  Mathematics: Trigonometry, 3rd March 25 minutes
  Mathematics: Trigonometry, 5th March 15 minutes
  Mathematics: Trigonometry, 3rd April 15 minutes
  ... and then at monthly intervals.

Chapter 1

Core 1 Pure Mathematics

The following topics are covered in this chapter:

- Algebra and functions
- Coordinate geometry in the (x, y) plane
- Sequences and series
- Differentiation
- Integration

Note: Calculators are *not* allowed in this module

1.1 Algebra and functions

After studying this section you should be able to:

- work with indices and surds
- use function notation
- solve quadratic equations and sketch the graphs of quadratic functions
- understand the definition of a polynomial
- recognise and sketch graphs of a range of functions
- use transformations to sketch graphs of related functions
- solve simultaneous equations – one linear and one quadratic
- solve linear and quadratic inequalities

LEARNING SUMMARY

Indices

You need to know these basic rules and be able to apply them.

$$a^m \times a^n = a^{m+n} \qquad a^m \div a^n = a^{m-n} \qquad (a^m)^n = a^{mn} \qquad a^{\frac{1}{n}} = \sqrt[n]{a}$$

$$a^{\frac{m}{n}} = (a^m)^{\frac{1}{n}} = (a^{\frac{1}{n}})^m \qquad a^{-n} = \frac{1}{a^n} \qquad (ab)^n = a^n b^n \qquad \left(\frac{a}{b}\right)^n = \frac{a^n}{b^n}$$

> Remember that, in general
> $(a+b)^n \neq a^n + b^n$
> and $(a-b)^n \neq a^n - b^n$.

Some important special cases are: $a^1 = a \qquad a^{\frac{1}{2}} = \sqrt{a}$

and $a^0 = 1 \qquad a^{-1} = \frac{1}{a}$ provided $a \neq 0$.

For example $9^{\frac{1}{2}} = \sqrt{9} = 3$ and $16^{\frac{3}{2}} = (16^{\frac{1}{2}})^3 = 4^3 = 64$.

Surds

A **surd** is the root of a whole number that has an **irrational** value.
Some examples are $\sqrt{2}$, $\sqrt{3}$ and $\sqrt{10}$.
You can often simplify a surd using the fact that $\sqrt{ab} = \sqrt{a} \times \sqrt{b}$ and choosing a or b to be a square number.

> An irrational number continues for ever after the decimal point without a repeating pattern.

For example $\sqrt{12} = \sqrt{4 \times 3} = \sqrt{4} \times \sqrt{3} = 2\sqrt{3}$ and $\sqrt{18} + 2\sqrt{32} = 3\sqrt{2} + 2 \times 4\sqrt{2} = 11\sqrt{2}$.

To simplify $\dfrac{3}{\sqrt{5}}$, multiply the numerator and the denominator by $\sqrt{5}$ to get

$\dfrac{3}{\sqrt{5}} \times \dfrac{\sqrt{5}}{\sqrt{5}} = \dfrac{3\sqrt{5}}{5}$. This is called **rationalising the denominator**.

Functions

A **function** may be thought of as a rule which takes each member x of a set and assigns, or **maps**, it to some value y known as its **image**.

x maps to y
y is the image of x.

$$x \longrightarrow \boxed{\text{Function}} \longrightarrow y$$

A letter such as f, g or h is often used to stand for a function. The function which squares a number and adds on 5, for example, can be written as $f(x) = x^2 + 5$. The same notation may also be used to show how a function affects particular values. For this function, $f(4) = 4^2 + 5 = 21$, $f(-10) = (-10)^2 + 5 = 105$ and so on.

f(x) is read as 'f of x'.

An alternative notation for the same function is $f: x \mapsto x^2 + 5$.

Quadratics

In algebra, any expression of the form $ax^2 + bx + c$ where $a \neq 0$ is called a **quadratic expression**.

You need to be able to expand brackets, **for example**

$$(2x + 3)(x - 4) = 2x^2 - 8x + 3x - 12$$
$$= 2x^2 - 5x - 12$$

It is useful to remember these special results:

$$(x + a)^2 = x^2 + 2ax + a^2 \qquad (x - a)^2 = x^2 - 2ax + a^2 \qquad (x + a)(x - a) = x^2 - a^2$$

Some examples are:

$$(x + 3)^2 = x^2 + 6x + 9 \qquad\qquad (x - 5)^2 = x^2 - 10x + 25$$
$$(x + 7)(x - 7) = x^2 - 49 \qquad\qquad (x + \sqrt{3})(x - \sqrt{3}) = x^2 - 3$$

To solve problems in algebra you need to develop your skills so that you can recognise how to apply the basic results.

In this fraction, the denominator $2 - \sqrt{3}$ is irrational.

This is an example of rationalising the denominator.

For example, the fraction $\dfrac{1}{2 - \sqrt{3}}$ may be simplified by multiplying the numerator and denominator by $2 + \sqrt{3}$ to make an equivalent fraction in which the denominator is rational.

$(2 - \sqrt{3})(2 + \sqrt{3})$
$= 2^2 - (\sqrt{3})^2$
$= 4 - 3.$

$$\frac{1}{2 - \sqrt{3}} = \frac{1}{2 - \sqrt{3}} \times \frac{2 + \sqrt{3}}{2 + \sqrt{3}}$$

$$= \frac{2 + \sqrt{3}}{4 - 3} \qquad \text{(The denominator is now rational)}$$

$$= 2 + \sqrt{3}$$

Quadratic equations

Equations of the form $ax^2 + bx + c = 0$ (where $a \neq 0$) are **quadratic equations**.

Some quadratic equations can be solved by **factorising** the quadratic expression.

Example Solve $2x^2 - x - 3 = 0$.

$$(2x - 3)(x + 1) = 0$$

Either $2x - 3 = 0$ or $x + 1 = 0$

so $x = \frac{3}{2}$ or $x = -1$.

Factorise.

Some equations can be converted into a quadratic equation by substitution.

Example Solve $x^{\frac{2}{3}} + x^{\frac{1}{3}} - 6 = 0$

Substituting $y = x^{\frac{1}{3}}$ gives

$$y^2 + y - 6 = 0$$

$$(y + 3)(y - 2) = 0$$

Either $y + 3 = 0$ or $y - 2 = 0$

so $y = -3$ or $y = 2$.

Re-writing in terms of x gives $x^{\frac{1}{3}} = -3$ or $x^{\frac{1}{3}} = 2$,

so $x = -27$ or $x = 8$.

If the quadratic will not factorise then you can try **completing the square**.

Example Solve $x^2 - 6x + 1 = 0$

$x^2 - 6x = (x - 3)^2 - 9$ so
$x^2 - 6x + 1 = (x - 3)^2 - 8$.

$(x - 3)^2 - 8 = 0$ (Now x only appears once in the equation)

$$(x - 3)^2 = 8$$

$$x - 3 = \pm \sqrt{8}$$

$\sqrt{8} = \sqrt{(4 \times 2)} = 2\sqrt{2}$.

$$x = 3 \pm 2\sqrt{2}.$$

(The solutions are $3 + 2\sqrt{2}$ and $3 - 2\sqrt{2}$)

The method shown for completing the square can be adapted for the general form of a quadratic. This gives the **quadratic formula**.

$$ax^2 + bx + c = 0$$

Multiply both sides by $4a$.

$$4a^2x^2 + 4abx + 4ac = 0$$

Note: $4a^2x^2 + 4abx + b^2 = (2ax + b)^2$.

Add b^2 to complete the square and subtract it again to keep things the same.

$$4a^2x^2 + 4abx + b^2 + 4ac - b^2 = 0$$

$$(2ax + b)^2 + 4ac - b^2 = 0$$

$$(2ax + b)^2 = b^2 - 4ac$$

Now x only appears once in the equation.

Find the square root of both sides.

$$2ax + b = \pm \sqrt{b^2 - 4ac}$$

Rearrange to find x.

$$x = \frac{-b \pm \sqrt{b^2 - 4ac}}{2a}$$

You must learn this formula.

Example Solve $3x^2 - 4x - 2 = 0$ giving your answers in surd form.

Comparing this equation with the general form gives $a = 3$, $b = -4$ and $c = -2$. Substitute this information into the formula:

In Core 1 you do not have access to a calculator so you will have to leave the answer in surd form.

$$x = \frac{4 \pm \sqrt{(-4)^2 - 4(3)(-2)}}{6} = \frac{4 \pm \sqrt{40}}{6} = \frac{4 \pm 2\sqrt{10}}{6} = \frac{2 \pm \sqrt{10}}{3}$$

so $x = \frac{2 + \sqrt{10}}{3}$ or $x = \frac{2 - \sqrt{10}}{3}$.

Discriminant

In the formula, the value of $b^2 - 4ac$ is called the **discriminant**. This value can be used to give information about the solutions without having to solve the equation.

$b^2 - 4ac > 0$ *two* distinct real solutions (the solutions are often called **roots**)

$b^2 - 4ac = 0$ *one* real solution (often thought of as a **repeated root**)

$b^2 - 4ac < 0$ *no* real solutions.

> In the previous example, $b^2 - 4ac = 40 > 0$ and two real roots were found.

The solutions of a quadratic equation correspond to where the graph of the quadratic function crosses the x-axis. There are three possible situations depending on the value of the discriminant.

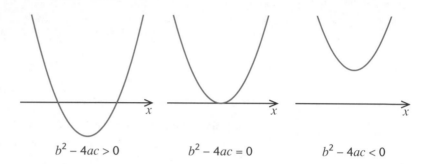

$b^2 - 4ac > 0$ $b^2 - 4ac = 0$ $b^2 - 4ac < 0$

> The diagrams correspond to the situation where $a > 0$. The same principle applies when $a < 0$ but the diagrams appear the other way up.

Example   The equation $5x^2 + 3x + p = 0$ has a repeated root. Find the value of p.

In this case, $a = 5$, $b = 3$ and $c = p$.

For a repeated root $b^2 - 4ac = 0$ so $9 - 20p = 0$, giving $p = 0.45$

Quadratic graphs

The graph of $y = ax^2 + bx + c$ is a parabola.

$a > 0$ $a < 0$

The methods used for solving quadratic equations can also be used to give information about the graphs.

Example   Sketch the graph of $y = x^2 - x - 6$.
              Find the coordinates of the lowest point on the curve.

The curve will cross the x-axis when $y = 0$. You can find these points by solving the equation $x^2 - x - 6 = 0$.

> The curve will cross the y-axis when $x = 0$, giving $y = -6$.

$x^2 - x - 6 = 0 \implies (x + 2)(x - 3) = 0$

$\implies x = -2$ or $x = 3$

The curve is symmetrical so the lowest point occurs mid-way between −2 and 3 and this is given by (−2 + 3) ÷ 2 = 0.5

When $x = 0.5$, $y = 0.5^2 - 0.5 - 6 = -6.25$

The lowest point on the curve is (0.5, −6.25).

The vertex of the curve occurs at its maximum or minimum point.

Completing the square gives information about the **vertex** of the curve even if the equation will not factorise.

Example   Find the coordinates of the vertex of the curve $y = x^2 + 2x + 3$.

You need to recognise that $x^2 + 2x + 1 = (x + 1)^2$,

then, completing the square, $x^2 + 2x + 3 = x^2 + 2x + 1 + 2 = (x + 1)^2 + 2$.

The equation of the curve can now be written as $y = (x + 1)^2 + 2$.

$(x + 1)^2$ cannot be negative so its minimum value is zero, when $x = -1$.

This means that the minimum value of y is 2 and this occurs when $x = -1$.

The vertex of the curve is at (−1, 2).

Polynomials

A quadratic is one example of a **polynomial**. In general, a polynomial takes the form:

$$a_n x^n + a_{n-1} x^{n-1} + a_{n-2} x^{n-2} + \dots + a_0,$$

This is much simpler than it looks.

where a_n, a_{n-1}, ... a_0 are constants and n is a positive whole number.

For example, $x^4 - 2x^3$ is a polynomial. In this case $a_4 = 1$, $a_3 = -2$ and a_2, a_1, a_0 are all zero.

The **degree** of a polynomial is the highest power of x that it includes, so the degree of $x^4 - 2x^3$ is 4. A quadratic is a polynomial of degree 2, a cubic is a polynomial of degree 3 and so on.

Higher order polynomials can also be factorised. **For example,** $x^3 + 6x^2 + 11x + 6 = (x + 1)(x + 2)(x + 3)$

More graphs

The graph of a cubic function $y = ax^3 + bx^2 + cx + d$ can take a number of forms.

$a > 0$ $a < 0$

The graph of a cubic function that can be factorised as $y = (x - p)(x - q)(x - r)$ will cross the x-axis at p, q and r. If any two of p, q and r are the same then the x-axis will be a tangent to the curve at that point.

KEY POINT

For example, the graph of $y = (x+2)(x-3)^2$ looks like this:

This shows a repeated root of $(x+2)(x-3)^2 = 0$ at $x = 3$.

The graph of $y = x^n$ where n is an integer has:
rotational symmetry about the **origin** when n is **odd**
reflective symmetry in the y-axis when n is **even**.

The graphs of $y = x^3$, $y = x^5$, $y = x^7$ ... look something like this:

The graphs of $y = x^2$, $y = x^4$, $y = x^6$ ... look something like this:

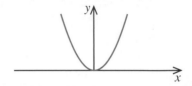

For higher powers of x the graphs are flatter between −1 and 1 and steeper elsewhere.

A straight line that a graph approaches ever more closely without actually touching it is called an **asymptote**.

The graphs of $y = x^{-1}$, $y = x^{-3}$, $y = x^{-5}$ ... look something like this:

The graphs of $y = x^{-2}$, $y = x^{-4}$, $y = x^{-6}$ ... look something like this:

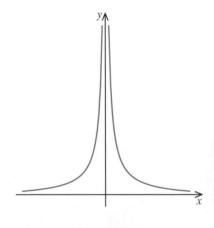

As the powers of x become more negative, the graphs become steeper between −1 and 1 and flatter elsewhere.

Both the x and y-axes are asymptotes for these graphs.

The x-axis and the positive y-axis are asymptotes for these graphs.

The graphs of $y = k\sqrt{x}$, $k > 0$, look something like this. These graphs do not have any asymptotes.

Transforming graphs

The graph of some new function can often be obtained from the graph of a known function by applying a transformation. A summary of the standard transformations is given in the table.

Known function	New function	Transformation
$y = f(x)$	$y = f(x) + a$	Translation through a units parallel to y-axis.
	$y = f(x - a)$	Translation through a units parallel to x-axis.
	$y = af(x)$	One-way stretch with scale factor a parallel to the y-axis.
	$y = f(ax)$	One-way stretch with scale factor $\frac{1}{a}$ parallel to the x-axis.

You may need to apply a combination of transformations in some cases.

Example   The diagram shows the graph of a function, $y = f(x)$ for $1 \leqslant x \leqslant 3$

Use the same axes to show:
(a) $y = f(x) + 1$
(b) $y = f(x + 1)$
(c) $y = 2f(x)$
(d) $y = f(2x)$

(a)

(b)

(c)

(d)
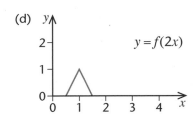

Simultaneous equations

The substitution method of solving pairs of linear simultaneous equations can also be applied when one of the equations is a quadratic.

Example   Find the coordinates of the points where the line $y = x + 1$ intersects the circle $x^2 + y^2 = 5$.

$$y = x + 1 \quad (1)$$
$$x^2 + y^2 = 5 \quad (2)$$

Substitute for y from (1) into (2): $x^2 + (x + 1)^2 = 5.$

Now expand the brackets: $x^2 + x^2 + 2x + 1 = 5.$

Arrange in the form $ax^2 + bx + c = 0$: $2x^2 + 2x - 4 = 0.$

Divide by 2. $x^2 + x - 2 = 0$

Solve the equation: $(x - 1)(x + 2) = 0$
Either $x = 1$ or $x = -2$

Substitute into (1) to find the y values: When $x = 1$, $y = 2$
When $x = -2$, $y = -1$

The coordinates of the points of intersection are (1, 2) and (−2, −1).

Core 1 Pure Mathematics

Geometrical interpretation of algebraic solutions

In the above example, when $y = x + 1$ and $x^2 + y^2 = 5$ are solved simultaneously, the resulting quadratic equation in x has two distinct solutions. This gives the two points of intersection of the line and the curve.

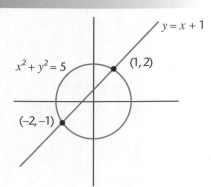

When you solve the equations of a line and a curve simultaneously and form a quadratic equation, $ax^2 + bx + c = 0$, the discriminant, $b^2 - 4ac$, gives information about the number of points of intersection.

If $b^2 - 4ac > 0$ the line and curve intersect in two distinct points.

If $b^2 - 4ac = 0$ the line is a tangent to the curve.

If $b^2 - 4ac < 0$ the line and the curve do not intersect.

Inequalities

Linear inequalities can be solved by rearrangement in much the same way as linear equations. However, care must be taken to reverse the direction of the inequality when multiplying or dividing by a negative.

Example   Solve the inequality $8 - 3x > 23$

Subtract 8 from both sides: $-3x > 15$
Divide both sides by -3: $x < -5$

An inequality which has x on both sides is treated like the corresponding equation.

Example   Solve the inequality $5x - 3 > 3x - 10$

Subtract $3x$ from both sides: $2x - 3 > -10$
Add 3 to both sides: $2x > -7$
Divide both sides by 2: $x > -3.5$

Quadratic inequalities are solved in a similar way to quadratic equations but a sketch graph is often helpful at the final stage.

Example   Solve the inequality $x^2 - 3x + 2 < 0$.

Factorise the quadratic expression: $(x - 1)(x - 2) < 0$.

Sketch the graph of $y = (x - 1)(x - 2)$:

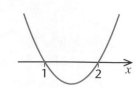

> $y = 0$ when $x = 1$ or when $x = 2$. The graph must cross the x-axis at these points.

> $(x - 1)(x - 2) < 0$ when the curve is below the x-axis.

The graph shows that $(x - 1)(x - 2) < 0$ for x values between 1 and 2.

It follows that $x^2 - 3x + 2 < 0$ when $1 < x < 2$.

20

If the quadratic expression cannot be factorised then the formula may be used to find the points of intersection of the curve with the x-axis.

Example Solve the inequality $x^2 + 2x - 5 > 0$.

$\sqrt{24} = \sqrt{4 \times 6} = 2\sqrt{6}$

Use the formula to solve $x^2 + 2x - 5 = 0$.
$$x = \frac{-2 \pm \sqrt{24}}{2} = \frac{-2 \pm 2\sqrt{6}}{2}$$

Simplify the result. $x = -1 \pm \sqrt{6}$

$x^2 + 2x - 5 > 0$ when the curve is above the x-axis.

Sketch the graph of $y = x^2 + 2x - 5$:

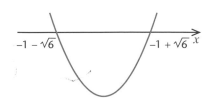

Write the solution as two separate inequalities.

From the sketch, $x^2 + 2x - 5 > 0$ when $x < -1 - \sqrt{6}$ or when $x > -1 + \sqrt{6}$.

Progress check

1 Simplify these surd expressions:

   (a) $\sqrt{72}$ (b) $(1 + \sqrt{3})(1 - \sqrt{3})$ (c) $5\sqrt{12} - 6\sqrt{3}$.

2 (a) Solve $2x^2 + x - 21 = 0$ by factorising.
   (b) Solve $x^2 - 6x + 7 = 0$ by completing the square.
       Give the roots in surd form.
   (c) Solve $3x^2 + 2x - 2 = 0$ using the quadratic formula.
       Give the roots in surd form.

3 Express $x^2 - 8x + 17$ in the form $(x - a)^2 + b$ and hence write down its minimum value.

4 Sketch the graph of $y = (x + 3)(x - 2)^2$.

5 The diagram shows the graph of $y = f(x)$ for $0 \leqslant x \leqslant 2$.
   Sketch the graph of $y = f(x - 2)$

6 Solve these equations simultaneously.
   $$x + 2y = 6$$
   $$4y^2 - 5x^2 = 36.$$

7 Find the values of A, B and C.
   $$3x^3 - 16x - 1 \equiv Ax(x - 1)(x + 2) + B(x - 1)^2 + C(4x^2 + 3).$$

1.2 Coordinate geometry in the (x, y) plane

After studying this section you should be able to:

- find the gradient of a line joining two points
- recognise equations of straight lines in various forms
- construct the equation of a straight line parallel to a given line and passing through a given point
- construct the equation of a straight line perpendicular to a given line and passing through a given point
- construct the equation of a straight line passing through two given points
- find the coordinates of the mid-point of two given points

Gradient of a line

The gradient of a line joining the points (x_1, y_1) and (x_2, y_2) is given by the formula $m = \dfrac{y_2 - y_1}{x_2 - x_1}$.

For example, the gradient of the line joining $(2, -5)$ and $(-1, 4)$ is
$$\frac{4 - (-5)}{-1 - 2} = \frac{9}{-3} = -3.$$

Straight lines

The general equation of a straight line is $ax + by + c = 0$. The equation of any straight line can be written in this form or as $y = mx + c$.

For example, the line $x + y = 5$ corresponds to $a = 1$, $b = 1$ and $c = -5$.

> The gradient of a vertical line is undefined.

Straight lines, apart from those parallel to the y-axis, can be written in the form $y = mx + c$. This is known as **gradient–intercept** form because the gradient (m) and the y-intercept (c) are clearly identified in the equation. This makes it easy to construct an equation when the gradient and intercept are known.

For example, the line with gradient 4 crossing the y-axis at -5 has equation $y = 4x - 5$.

Straight lines that are **parallel** must have the **same gradient**.

Example Find the equation of the straight line parallel to $y = 3x - 5$ and passing through the point (4, 2).

> This equation must be satisfied at the point where $x = 4$ and $y = 2$.

The line must have gradient 3 and so it can be written in the form $y = 3x + c$. Substituting $x = 4$ and $y = 2$ gives $2 = 3 \times 4 + c \Rightarrow c = -10$. The required equation is $y = 3x - 10$.

This result is used frequently to find the equation of a tangent or a normal to a curve at a given point.

The equation of a straight line with gradient m and passing through the point (x_1, y_1) can be written as $y - y_1 = m(x - x_1)$.

Using this in the example above gives $y - 2 = 3(x - 4)$. The equation is acceptable in this form but it can be rearranged to give $y = 3x - 10$ as before.

Example Find the equation of the straight line passing through the points $(-1, 5)$ and $(3, -2)$.

One approach is to use the form $y = mx + c$ to produce a pair of simultaneous equations:

Substituting $x = -1$ and $y = 5$ gives $\qquad\qquad 5 = -m + c$ $\qquad\qquad$ (1)
Substituting $x = 3$ and $y = -2$ gives $\qquad\qquad -2 = 3m + c$ $\qquad\qquad$ (2)

$(2) - (1)$ gives: $\qquad\qquad\qquad\qquad\qquad -7 = 4m \Rightarrow m = \dfrac{-7}{4}$.

Substituting for m in (1) gives: $\qquad\qquad 5 = \dfrac{7}{4} + c \Rightarrow c = \dfrac{13}{4}$.

The equation of the line is $y = \dfrac{-7}{4}x + \dfrac{13}{4}$. This is the same as $4y + 7x = 13$.

An alternative approach is to find the gradient directly and then use the form $y - y_1 = m(x - x_1)$.

Taking $(-1, 5)$ as (x_1, y_1) and $(3, -2)$ as (x_2, y_2),

$$m = \frac{-2 - 5}{3 - (-1)} = \frac{-7}{4} = -\frac{7}{4}.$$

You can use either of the points $(-1, 5)$ or $(3, -2)$.

The equation of the line is $y - 5 = -\dfrac{7}{4}(x + 1)$.

To show that this is the same as the previous result (you don't *need* to do this):

Multiply both sides of the equation by 4: $\qquad\qquad 4y - 20 = -7(x + 1)$.
Expand the brackets: $\qquad\qquad\qquad\qquad\qquad 4y - 20 = -7x - 7$.
Rearrange to give the previous result: $\qquad\qquad 4y + 7x = 13$.

An equation of the form $ax + by + c = 0$ can be rearranged into gradient–intercept form provided that $b \neq 0$. This becomes $y = -\frac{a}{b}x - \frac{c}{b}$ and shows that parallel lines may be produced by keeping a and b fixed and allowing c to change.

Note that $4x - 6y = 7$ is equivalent to $2x - 3y = 3.5$

For example, the lines $2x - 3y = 5$, $2x - 3y = -2$, $4x - 6y = 7$ are all parallel.

$m_1 \times m_2 = -1$.

When two straight lines are **perpendicular**, the **product** of their gradients is -1.

Example Find the equation of the straight line perpendicular to the line $4x + 3y = 12$ and passing through the point $(2, 5)$.

Rearrange the equation into gradient–intercept form: $\qquad y = -\frac{4}{3}x + 4$.
If the gradient of the required line is m then: $\qquad m \times -\frac{4}{3} = -1 \Rightarrow m = \frac{3}{4}$
Using the form $y - y_1 = m(x - x_1)$ gives: $\qquad y - 5 = \frac{3}{4}(x - 2)$.

If P has coordinates (x_1, y_1) and Q has coordinates (x_2, y_2) then the mid-point of PQ has coordinates $\left(\dfrac{x_1 + x_2}{2}, \dfrac{y_1 + y_2}{2}\right)$.

For example, the mid-point of the line joining $(-3, 1)$ and $(5, 7)$ is $\left(\dfrac{-3 + 5}{2}, \dfrac{1 + 7}{2}\right) = (1, 4)$.

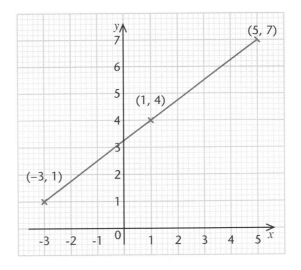

Progress check

1 The equation of a straight line is $y = 2x - 7$.

(a) Write down the equation of a parallel line crossing the y-axis at 5.
(b) Find the equation of a parallel line passing through the point $(4, 9)$.

2 Find the equation of the straight line joining the points $(-2, -9)$ and $(1, 3)$.

3 The equation of a straight line is $3x + 2y = 11$.

(a) Find the equation of a parallel line passing through the point $(5, 4)$.
(b) Find the equation of a perpendicular line passing through the point $(-2, 6)$.

4 Find the coordinates of the mid-point of $(-9, 3)$ and $(-5, 1)$.

5 P is the point $(-1, -4)$ and Q is the point $(5, -1)$. Find the equation of the line perpendicular to PQ and passing through its mid-point.

> The parallel line can be written in the form $3x + 2y = C$. To find C just substitute the coordinates $(5, 4)$ for x and y.

5 $y = -2x + 1.5$
4 $(-7, 2)$
3 (a) $3x + 2y = 23$ (b) $y - 6 = \frac{3}{2}(x + 2)$ or $3y - 2x - 22 = 0$.
2 $y = 4x - 1$
1 (a) $y = 2x + 5$ (b) $y = 2x + 1$

1.3 Sequences and series

After studying this section you should be able to:

- understand the different types of sequence, including the notation and formulae used to describe them
- recognise arithmetic progressions and calculate the sum of their series

Sequences and series

A list of numbers in a particular order, that follow some rule for finding later values, is called a **sequence**. Each number in a sequence is called a **term**, and terms are often denoted by $u_1, u_2, u_3, ..., u_n, ...$

One way to define a sequence is to give a formula for the nth term such as $u_n = n^2$. Substituting the values $n = 1, 2, 3, 4, ...$ produces the sequence $1, 4, 9, 16, ...$ and the value of any particular term can be calculated by substituting its position number into the formula. For example, in this case, the 50th term is $u_{50} = 50^2 = 2500$.

Another way to define a sequence is to give a starting value together with a rule that shows the connection between successive terms. This is sometimes called a **recursive definition**. For example $u_1 = 5$ and $u_{n+1} = 2u_n$ defines the sequence $5, 10, 20, 40, ...$. The rule $u_{n+1} = 2u_n$ is an example of a **recurrence relation**.

Two special sequences are the **arithmetic progression** (A.P.) and the **geometric progression** (G.P.). Geometric progression is studied in Core 2.

In an A.P. successive terms have a **common difference**, e.g. $1, 4, 7, 10, ...$. The first term is denoted by a and the common difference is d. With this notation, the definition of an A.P. may now be given as $u_1 = a$, $u_{n+1} = u_n + d$.

The terms of an A.P. take the form $a, a + d, a + 2d, a + 3d, ...$ and the nth term is given by $u_n = a + (n - 1)d.$.

A **series** is formed by adding together the terms of a sequence. The use of sigma notation can greatly simplify the way that series are written. For example, the series $1^2 + 2^2 + 3^2 + ... + n^2$ may be written as $\sum_{i=1}^{n} i^2$. The sum of the first n terms of a series is often denoted by S_n and so $S_n = u_1 + u_2 + u_3 + ... + u_n = \sum_{i=1}^{n} u_i$.

Arithmetic series

The **sum of an arithmetic series** is given by
$$S_n = a + (a + d) + (a + 2d) + ... + (a + (n - 1)d).$$

This may also be written as
$$S_n = l + (l - d) + (l - 2d) + ... + (l - (n - 1)d),$$
where l is the last term.

Adding the two versions gives
$$2S_n = (a + l) + (a + l) + (a + l) + ... (a + l)$$
$$= n(a + l).$$
$$S_n = \frac{n}{2}(a + l).$$

You need to know how to establish the general result.

> **KEY POINT**
>
> The sum of the first n terms of an arithmetic series is: $S_n = \dfrac{n}{2}(a+l)$.
>
> Substituting $l = a + (n-1)d$ gives the alternative form of the result:
>
> $$S_n = \dfrac{n}{2}(2a + (n-1)d).$$

A special case is the sum of the first n natural numbers:

$$1 + 2 + 3 + \dots + n = \dfrac{n}{2}(n+1).$$

For example, the sum of the first 100 natural numbers is $\dfrac{100}{2}(100 + 1) = 5050$.

When finding the sum of an A.P. you need to select the most appropriate version of the formula to suit the information.

Example   Find the sum of the first 50 terms of the series $15 + 18 + 21 + 24 + \dots$.
In this series, $a = 15$, $d = 3$ and $n = 50$.

Using $S_n = \dfrac{n}{2}(2a + (n-1)d)$ gives $S_{50} = \dfrac{50}{2}(2 \times 15 + 49 \times 3) = 4425$.

Progress check

1   Write down the first five terms of the sequence given by:
(a) $u_n = 2n^2$ (b) $u_1 = 10$, $u_{n+1} = 3u_n + 2$.

2   Find the 20th term of an A.P. with first term 7 and common difference 5.

3   Find the sum of the first 1000 natural numbers.

3 500 500
2 102
1 (a) 2, 8, 18, 32, 50   (b) 10, 32, 98, 296, 890

1.4 Differentiation

After studying this section you should be able to:

- differentiate functions of the form x^n
- use differentiation to find the exact value of the gradient of a curve
- find the equation of the tangent and normal to a curve at a given point

LEARNING SUMMARY

Gradient function

The **gradient** of a curve changes continuously along its length. Its value at any point P is given by the gradient of the **tangent** to the curve at P. The gradient may be found approximately by drawing.

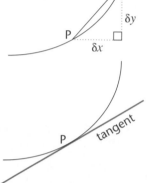

The exact value of the gradient is found by **differentiation**. This is the **limit** of the gradient of PQ as Q moves towards P.

Gradient of curve at P $\approx \dfrac{\delta y}{\delta x}$.

As Q moves towards P, $\delta x \to 0$ and $\dfrac{\delta y}{\delta x} \to \dfrac{\mathrm{d}y}{\mathrm{d}x}$.

> $\dfrac{\mathrm{d}y}{\mathrm{d}x}$ is the **gradient function** and represents the **derivative** of y with respect to x.
>
> The graph of $y = kx^n$ has gradient function $\dfrac{\mathrm{d}y}{\mathrm{d}x} = nkx^{n-1}$.
>
> KEY POINT

Example Find the gradient of the curve $y = 3x^2$ at the point P (5, 75).

$$\frac{\mathrm{d}y}{\mathrm{d}x} = 2 \times 3x^1 = 6x. \quad \text{At P, } x = 5 \text{ so the gradient is } 6 \times 5 = 30.$$

Function notation may also be used for derivatives. If $y = f(x)$ then $\dfrac{\mathrm{d}y}{\mathrm{d}x} = f'(x)$.

The notation is useful for stating some of the rules of differentiation.

> If $y = f(x) \pm g(x)$ then $\dfrac{\mathrm{d}y}{\mathrm{d}x} = f'(x) \pm g'(x)$
>
> KEY POINT

Example $f(x) = x^3 + 5x + 2$. Find (a) $f'(x)$ (b) $f'(4)$.

Differentiate each term separately.
Remember that $x^0 = 1$.
When you differentiate a constant, the result is zero.

(a) $f'(x) = 3x^2 + 5$ (b) $f'(4) = 3 \times 4^2 + 5 = 53$.

You often need to express a function in the right form before you can differentiate it.

Example Differentiate: (a) $f(x) = \sqrt{x}$ (b) $f(x) = (x - 5)(x + 3)$ (c) $f(x) = \dfrac{x^3 + 1}{x^2}$.

(a) $\sqrt{x}$ has to be written as a power of x to use the rule $f'(x) = nkx^{n-1}$.

$\sqrt{x} = x^{\frac{1}{2}}$ so $f(x) = x^{\frac{1}{2}}$ and $f'(x) = \frac{1}{2}x^{-\frac{1}{2}}$.

> You cannot differentiate a product term by term.

(b) The brackets must first be removed and *then* you can differentiate term by term.

$f(x) = (x-5)(x+3) = x^2 - 2x - 15$.

$f'(x) = 2x - 2$.

> You cannot differentiate a quotient term by term.

(c) Divide $x^3 + 1$ by x^2 first giving $f(x) = x + x^{-2}$.

Now, $f'(x) = 1 - 2x^{-3} = 1 - \dfrac{2}{x^3}$.

You also have to differentiate expressions when n is not a positive integer. You may have to change the expression into index form first.

Example 1 Differentiate $y = x^{\frac{3}{2}} + \dfrac{3}{x^2}$

Re-write as $y = x^{\frac{3}{2}} + 3x^{-2}$ and differentiate term by term to give

$$\frac{dy}{dx} = \frac{3}{2}x^{\frac{1}{2}} + (-2) \times 3x^{-3}$$

$$= \frac{3}{2}\sqrt{x} - \frac{6}{x^3}$$

Example 2 Differentiate $y = x\sqrt{x}$

Re-write as $y = x^{\frac{3}{2}}$ giving

$$\frac{dy}{dx} = \frac{3}{2}\sqrt{x}$$

Tangents and normals

You can find the gradient of the tangent to a curve at a point by differentiation. Then you can use the techniques described in the Coordinate Geometry section to find the **equation of the tangent** and the **equation of the normal** at the given point.

Example Find the equation of the tangent and the normal to the curve $y = x^3 - 4x$ at the point (2, 0).

Differentiate the equation of the curve to give $\dfrac{dy}{dx} = 3x^2 - 4$.

The gradient of the tangent at (2, 0) is $3 \times 2^2 - 4 = 8$.
Using $y - y_1 = m(x - x_1)$ gives the equation of the tangent as $y = 8(x - 2)$.

> The normal at a point is perpendicular to the tangent.

The gradient of the normal is $-\frac{1}{8}$. Using $y - y_1 = m(x - x_1)$ again gives the equation of the normal as $y = -\frac{1}{8}(x - 2)$. You could rearrange this to give $x + 8y = 2$.

Progress check

1 Differentiate with respect to x:

$y = 6\sqrt{x}$.

2 Differentiate with respect to x:

$f(x) = \dfrac{2x+1}{x^3}$.

3 Find the gradient of the curve $y = (4x-1)^2$ at the point (1, 9).

4 Find the equation of: (a) the tangent   (b) the normal
to the curve $y = x^3 - 2x$ at the point (2, 4).

5 Find $\dfrac{\mathrm{d}y}{\mathrm{d}x}$ in each of the following questions.

(a) $y = x^{\frac{5}{3}} - \dfrac{1}{x}$

(b) $y = x^3\sqrt{x}$

5 (a) $\frac{5}{3}x^{\frac{2}{3}} + \dfrac{1}{x^2}$

(b) $\frac{7}{2}x^{\frac{5}{2}}$

4 (a) Tangent: $y = 10x - 16$;   (b) normal: $x + 10y - 42 = 0$.

3 24

2 $\dfrac{3}{x^3} - \dfrac{4}{x^4}$

1 $\dfrac{3}{\sqrt{x}}$

1.5 Integration

After studying this section you should be able to:

- find an indefinite integral and understand what it represents

Indefinite integration

The idea of a **reverse process** is an important one in many areas of mathematics. In **calculus**, the reverse process of differentiation is **integration** and this turns out to be an extremely important process in its own right, with many powerful applications.

When you differentiate x^n with respect to x you may think of the process involving two stages:

$$x^n \rightarrow \boxed{\text{multiply by the power}} \rightarrow \boxed{\text{reduce the power by 1}} \rightarrow nx^{n-1}$$

This *suggests* that the reverse process is given by:

$$\frac{x^{n+1}}{n+1} \leftarrow \boxed{\text{divide by the power}} \leftarrow \boxed{\text{increase the power by 1}} \leftarrow x^n$$

However, this doesn't give the *complete* picture. The reason is that you can differentiate $x^n +$ (any constant) and still obtain nx^{n-1}. You need to take this into account when you reverse the process.

$\int$ is the symbol for integration and dx is used to show that the integration is with respect to the variable x.

> **KEY POINT**
>
> The result is written as $\displaystyle\int x^n\,dx = \frac{x^{n+1}}{n+1} + c$ where c is called the **constant of integration**. Note that $n \neq -1$.

Some examples are:

Notice that $\dfrac{1}{\left(\frac{3}{2}\right)} = \dfrac{2}{3}$.

$$\int x^2\,dx = \frac{x^3}{3} + c \qquad \int \sqrt{x}\,dx = \int x^{\frac{1}{2}}\,dx = \frac{2}{3}x^{\frac{3}{2}} + c \qquad \int 3\,dx = 3x + c$$

$$\int \frac{1}{x^2}\,dx = \int x^{-2}\,dx = -x^{-1} + c = -\frac{1}{x} + c.$$

Sums and differences of functions are treated in the same way as in differentiation, by dealing with each term separately. The general rules are given below.

Function notation is useful here but you don't need to state the rules every time you use them.

> **KEY POINT**
>
> $$\int (f(x) \pm g(x))\,dx = \int f(x)\,dx \pm \int g(x)\,dx$$
>
> $$\int af(x)\,dx = a\int f(x)\,dx \quad \text{(where } a \text{ is a constant)}$$

For example,

$$\int (3x^2 - 5x + 2)\,dx = x^3 - \tfrac{5}{2}x^2 + 2x + c.$$

In some situations you are given enough information to find the value of the constant.

Example Find the equation of the curve with gradient function $3x^2$ passing through $(2, 5)$.

Using the point $(2, 5)$ gives
$5 = 2^3 + c$
so $c = -3$.

$$\frac{dy}{dx} = 3x^2 \Rightarrow y = \int 3x^2 \, dx \Rightarrow y = x^3 + c. \quad \text{When } x = 2, y = 5 \Rightarrow c = -3.$$

The equation of the curve is $y = x^3 - 3$.

Progress check

1 Integrate each of these with respect to x.

(a) $\dfrac{1}{\sqrt{x}}$

(b) $\sqrt{x^3}$

2 A curve has gradient function $4x^3 + 1$ and passes through the point $(1, 9)$. Find the equation of the curve.

2 $y = x^4 + x + 7$

1 (a) $2\sqrt{x} + c$ (b) $\frac{2}{5}x^{5/2} + c$

Sample questions and model answers

'Write down' means that you don't need to show any working to get the marks. But you might still find it useful as a way to clarify your thoughts.

Remember that $a^{-n} = \dfrac{1}{a^n}$

1

Write down the exact value of 3^{-3}.

(a) $3^{-3} = \dfrac{1}{3^3} = \dfrac{1}{27}$.

2

(a) Write $2x^2 - 12x + 11$ in the form $a(x+b)^2 + c$.

Start by taking out the factor of 2 common to the first two terms.

(b) State the minimum value of $2x^2 - 12x + 11$ and give the value of x where this occurs.

Keep the 11 separate rather than have a fraction inside the brackets.

(c) Solve the equation $2x^2 - 12x + 11 = 0$ and express your answer in surd form.

(d) Sketch the graph of $y = 2x^2 - 12x + 11$.

Complete the square and simplify.

(a) $2x^2 - 12x + 11 = 2(x^2 - 6x) + 11$

$$= 2((x-3)^2 - 9) + 11$$

$$= 2(x-3)^2 - 7.$$

The minimum value occurs when the expression in the brackets equals zero.

(b) The minimum value is -7 and this occurs when $x = 3$.

(c) $2x^2 - 12x + 11 = 0$

$$\Rightarrow 2(x-3)^2 - 7 = 0$$

$$\Rightarrow (x-3)^2 = \tfrac{7}{2} = \tfrac{14}{4}$$

$$\Rightarrow x - 3 = \pm \tfrac{1}{2}\sqrt{14}$$

$$\Rightarrow x = 3 \pm \tfrac{1}{2}\sqrt{14}.$$

(d)

When $x = 0$, $y = 11$.

Use the information you have found to position the curve and label the key points.

Sample questions and model answers (continued)

3

A curve has equation $y = (2 - x)(2 + x)$.

The point P(−1, 3) lies on the curve.

(a) Find the gradient of the tangent to the curve at P.

(b) Find the equation of the normal to the curve at P in the form $px + qy = r$.

(c) The normal to the curve at P crosses the y-axis at A and the x-axis at B. Find the area of triangle AOB.

(a) $y = (2 - x)(2 + x)$

Expand the brackets before differentiating.

$\quad = 4 - x^2$

$\dfrac{dy}{dx} = -2x.$

When $x = -1$, $\dfrac{dy}{dx} = -2(-1) = 2$

The gradient of the tangent at P is 2.

(b) The gradient of the normal at P is $-\frac{1}{2}$.

The product of the gradients must be −1, as they are at right angles to each other.

Equation of the normal at P:

This is using the equation of a straight line in the form $y - y_1 = m(x - x_1)$.

$y - 3 = -\frac{1}{2}(x - (-1))$

$y - 3 = -\frac{1}{2}(x + 1)$

$2(y - 3) = -(x + 1)$

$2y - 6 = -x - 1$

$x + 2y = 5$

Write the equation in the required form.

(c) When $x = 0$, $y = \frac{5}{2} \Rightarrow$ A is the point $(0, \frac{5}{2})$

Find where the line intersects the axes.

When $y = 0$, $x = 5 \Rightarrow$ B is the point $(5, 0)$

Area of triangle AOB $= \frac{1}{2} \times 5 \times \frac{5}{2} = \frac{25}{4} = 6\frac{1}{4}$ units^2.

4

A curve has gradient function $2x - 5$. The point P(2, −1) lies on the curve.

Find the equation of the curve.

$\dfrac{dy}{dx} = 2x - 5$

$y = \displaystyle\int (2x - 5)\ dx$

Remember the integration constant c here.

$\quad = x^2 - 5x + c$

When $x = 2$, $y = -1$

Use the fact that P lies on the curve to find the value of c.

$\Rightarrow -1 = 2^2 - 5(2) + c$

$c = -1 - 4 + 10 = 5$

The curve has equation $y = x^2 - 5x + 5$.

Practice examination questions

1 Write down the exact value of $49^{-\frac{1}{2}}$.

2 Given that $8^{x-3} = 4^{x+1}$, find the value of x.

3 (a) Write $x^2 - 4x + 1$ in the form $(x + a)^2 + b$.

   (b) Solve $x^2 - 4x + 1 = 0$ and express your answer in surd form.

   (c) State the coordinates of the lowest point on the graph of $y = x^2 - 4x + 1$.

   (d) Sketch the graph.

4 A function $f(x)$ is defined by:

   $f(x) = x^3 + 4x^2 - 3x - 18$

   for all real values of x.

   (a) Sketch the graph of $y = f(x)$.

   (b) Sketch the graphs of $y = f(x) + 10$ and $y = f(x - 2)$.

5 Find the coordinates of the points A and B at the intersection of these graphs.

6 Solve the inequality

   $3x^2 - 8x - 7 < 2x^2 - 3x - 11$.

Practice examination questions *(continued)*

7 (a) The diagram shows a straight line passing through the points A(−1,−5) and B(4, 1).
   Find its equation in the form $ax + by + c = 0$.

   (b) A second line l, passes through the mid-point of A and B at right angles to AB.

   Find its equation in the form $ax + by + c = 0$.

   (c) The line l crosses the y-axis at C. Find the coordinates of C.

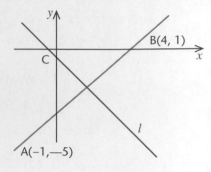

8 The diagram shows two lines l and m at right angles to each other. The equation of the line m is $3x + 2y = 6$. The line l passes through the point with coordinates (5, 4).
   l and m cross the x-axis at A and B respectively.

   (a) Find the equation of the line l.

   (b) Show that AB = 3.

9 The 10th term of an arithmetic progression is 74 and the sum of the first 20 terms is 1510. Find the first term and the common difference.

10 The diagram shows a straight line crossing the curve $y = (x − 1)^2 + 4$ at the points P and Q. It passes through (0, 7) and (7, 0).

   (a) Write down the equation of the straight line through P and Q.

   (b) Find the coordinates of P and Q.

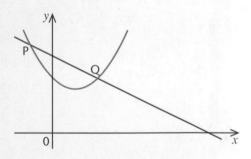

Chapter 2

Core 2 Pure Mathematics

The following topics are covered in this chapter:

- *Algebra and functions*
- *Coordinate geometry in the (x, y) plane*
- *Sequences and series*
- *Trigonometry*

- *Exponentials and logarithms*
- *Differentiation*
- *Integration*

2.1 Algebra and functions

After studying this section you should be able to:

- *carry out algebraic division*
- *use the remainder and factor theorems*

LEARNING SUMMARY

Algebraic division

If you work out $27 \div 4$ for example, then you obtain 6 as the **quotient** and 3 as the **remainder**. You can write $27 = 4 \times 6 + 3$ to show the connection between these values.

The same thing applies in algebra when one polynomial is divided by another.

Example   Find the quotient and remainder when $x^2 + 7x - 5$ is divided by $x - 4$.

Using $ax + b$ for the quotient and c for the remainder gives:

$$x^2 + 7x - 5 \equiv (x - 4)(ax + b) + c.$$
$$\equiv ax^2 + (b - 4a)x + c - 4b.$$

Equating coefficients of x^2 gives: $a = 1$.

Equating coefficients of x gives: $b - 4a = 7$

$$\Rightarrow b - 4 = 7 \qquad \text{(using the result } a = 1 \text{ from above)}$$
$$\Rightarrow b = 11.$$

Equating the constant terms gives: $c - 4b = -5$

$$\Rightarrow c - 44 = -5 \quad \text{(using the result } b = 11 \text{ from above)}$$
$$\Rightarrow c = 39.$$

Substituting for a, b and c in [1] gives $x^2 + 7x - 5 \equiv (x - 4)(x + 11) + 39$.

The remainder theorem

In the previous example, the quotient and remainder of $(x^2 + 7x - 5) \div (x - 4)$ were found by using $x^2 + 7x - 5 \equiv (x - 4)(ax + b) + c$ and equating coefficients.

Another way to use this identity is to substitute particular values for x. Notice that when $x = 4$, the RHS simplifies to c and so the remainder is easily found by substituting $x = 4$ on the LHS. This gives $c = 4^2 + 7 \times 4 - 5 = 39$ as before.

When $x = 4$, $(x - 4) = 0$ so $(x - 4)(ax + b) = 0$ and the RHS simplifies to c.

In its generalised form this result is known as the **remainder theorem**, which states:

> When a polynomial $f(x)$ is divided by $(x-a)$ the remainder is $f(a)$.
>
> KEY POINT

Example Find the remainder when the polynomial $f(x) = x^3 + x - 5$ is divided by:

(a) $(x-3)$ (b) $(x+2)$ (c) $(2x-1)$

> Choose x so that $x - 3 = 0$.

(a) The remainder is $f(3) = 3^3 + 3 - 5 = 25$

(b) The remainder is $f(-2) = (-2)^3 - 2 - 5 = -15$

> Choose x so that $2x - 1 = 0$.

(c) The remainder is $f(0.5) = 0.5^3 + 0.5 - 5 = -4.375$

The factor theorem

A special case of the remainder theorem occurs when the remainder is 0. This result is known as the **factor theorem**, which states:

> If $f(x)$ is a polynomial and $f(a) = 0$ then $(x-a)$ is a factor of $f(x)$.
>
> KEY POINT

Example Show that $(x+3)$ is a factor of $x^3 + 5x^2 + 5x - 3$.

$$\text{Taking } f(x) = x^3 + 5x^2 + 5x - 3$$

$$f(-3) = (-3)^3 + 5(-3)^2 + 5(-3) - 3$$

$$= -27 + 45 - 15 - 3 = 0.$$

By the factor theorem $(x+3)$ is a factor of $x^3 + 5x^2 + 5x - 3$.

Example A polynomial is given by $f(x) = 2x^3 + 13x^2 + 13x - 10$.

(a) Find the value of $f(2)$ and $f(-2)$.

(b) State one of the factors of $f(x)$.

(c) Factorise $f(x)$ completely.

(a) $f(2) = 2(2^3) + 13(2^2) + 13(2) - 10 = 84$

$f(-2) = 2(-2)^3 + 13(-2)^2 + 13(-2) - 10 = 0.$

(b) Since $f(-2) = 0$, $(x+2)$ is a factor of $f(x)$ by the factor theorem.

> $f(x)$ is a cubic so the linear factor $(x+2)$ must be combined with a quadratic.

(c) $2x^3 + 13x^2 + 13x - 10 \equiv (x+2)(ax^2 + bx + c)$.

This is an **identity** and so the values of a, b and c may be found by comparing coefficients.

Comparing the x^3 terms gives:	$a = 2$.
Comparing the x^2 terms gives:	$13 = 2a + b \Rightarrow b = 9$.
Comparing the constant terms gives:	$c = -5$.
It follows that:	$f(x) = (x+2)(2x^2 + 9x - 5)$.

Factorising the quadratic part in the usual way gives $f(x) = (x+2)(2x-1)(x+5)$.

Progress check

1 Find the remainder when $x^3 + 2x^2 - 5x + 1$ is divided by $(x - 2)$.
2 Given that f(x) = $2x^3 + 7x^2 - 3x - 18$ and that $x = -2$ is a root of the equation
   f(x) = 0, factorise f(x) completely.

2 (x + 2) (x + 3) (2x − 3)

1 7

2.2 Coordinate geometry in the (x, y) plane

After studying this section you should be able to:

- *use the equation of a circle and circle properties*

LEARNING SUMMARY

The equation of a circle

> **KEY POINT**
>
> The equation of a circle with centre (a, b) and radius r is
> $(x - a)^2 + (y - b)^2 = r^2$.

If the centre is at (0, 0), the equation is $x^2 + y^2 = r^2$

This result is based on Pythagoras' theorem.

$(x - a)^2 + (y - b)^2 = r^2$

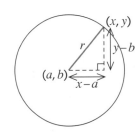

For example, the equation of a circle with centre (4, 0) and radius 5 units is
$(x - 4)^2 + y^2 = 25$.

> **KEY POINT**
>
> An alternative form of the equation of a circle is $x^2 + y^2 + 2gx + 2fy + c = 0$.
> The centre of the circle is $(-g, -f)$ and the radius is $\sqrt{g^2 + f^2 - c}$.

For example, the circle with equation $x^2 + y^2 + 6x - 10y + 18 = 0$ has centre $(-3, 5)$
and radius $\sqrt{(-3)^2 + 5^2 - 18} = \sqrt{16} = 4$ units.

You can also show this by completing the square on the x terms and on the y
terms as follows:

$$x^2 + 6x + y^2 - 10y + 18 = 0$$
$$(x + 3)^2 - 9 + (y - 5)^2 - 25 + 18 = 0$$
$$(x + 3)^2 + (y - 5)^2 = 16$$

Compare with
$(x - a)^2 + (y - b)^2 = r^2$.

The centre is at $(-3, 5)$ and the radius is 4.

Circle properties

You need to remember and be able to use these **circle properties**.

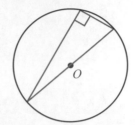

The angle in a **semicircle** is a right angle.

The perpendicular from the centre of a circle to a **chord** bisects the chord.

In the diagram, $AX = XB$.

The **tangent** to a circle is perpendicular to the radius at its point of contact.

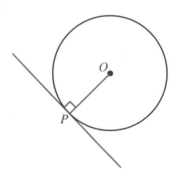

The normal to the circle at P is along OP, as it is perpendicular to the tangent.

If you know the gradient of OP, you can work out the gradient of the tangent using the fact that for perpendicular lines, the product of the gradients is -1.

Progress check

1 (a) Find the equation of a circle with centre $(-3, 8)$ and radius 5 units.
   (b) A circle has equation $x^2 + y^2 - 12x + 8y + 43 = 0$.
        Find its centre and radius.

1 (a) $(x + 3)^2 + (y - 8)^2 = 25$ (b) $(6, -4)$, 3 units

2.3 Sequences and series

After studying this section you should be able to:

- understand the different types of sequence, including the notation and formulae used to describe them
- recognise geometric progressions and calculate the sum of their series
- determine whether a geometric series with an infinite number of terms has a finite sum and calculate its value when it exists
- use the binomial expansion for positive integer powers of n.

LEARNING SUMMARY

Geometric series

Two special sequences are the **arithmetic progression** (A.P.), which was covered in Core 1, and the **geometric progression** (G.P.).

In a G.P. successive terms are connected by a **common ratio**, e.g. 3, 6, 12, 24, … . The definition of a G.P. may be given as $u_1 = a$, $u_{n+1} = ru_n$ where r is the common ratio.

The terms of a G.P. take the form $a, ar, ar^2, ar^3, …$ and the nth term is given by $u_n = ar^{n-1}$.

A **series** is formed by adding together the terms of a sequence. The use of sigma notation can greatly simplify the way that series are written. For example, the series $1^2 + 2^2 + 3^2 + … + n^2$ may be written as $\sum_{i=1}^{n} i^2$. The sum of the first n terms of a series is often denoted by S_n and so $S_n = u_1 + u_2 + u_3 + … + u_n = \sum_{i=1}^{n} u_i$.

The sum of a G.P. is given by $\qquad S_n = a + ar + ar^2 + ar^3 + … + ar^{n-1}$.

Multiplying throughout by r gives $rS_n = ar + ar^2 + ar^3 + … + ar^{n-1} + ar^n$.

Subtracting gives: $\qquad\qquad S_n - rS_n = a - ar^n$

$$\Rightarrow S_n(1 - r) = a(1 - r^n)$$

> You need to know how to establish this result.

> So the sum of the first n terms of a geometric series is $S_n = \dfrac{a(1 - r^n)}{1 - r}$.
>
> **KEY POINT**

If $r > 1$ then it can be convenient to use the result in the form $S_n = \dfrac{a(r^n - 1)}{r - 1}$.

Example Find the sum of the first 20 terms of the series $8 + 12 + 18 + 27 + …$ to the nearest whole number.

In this series, $a = 8$, $r = 1.5$ and $n = 20$.

This gives $S_{20} = \dfrac{8(1.5^{20} - 1)}{1.5 - 1} = 53188.107… = 53188$ to the nearest whole number.

Provided that $|r| < 1$, the sum of a geometric series converges to $\dfrac{a}{1-r}$ as n tends to infinity. This is known as the **sum to infinity** of a geometric series.

For example, $1 + \dfrac{1}{2} + \dfrac{1}{4} + \dfrac{1}{8} + \ldots = \dfrac{1}{1-\frac{1}{2}} = 2.$

Binomial expansion

An expression which has two terms, such as $a+b$ is called a **binomial**. The expansion of something of the form $(a+b)^n$ is called a **binomial expansion**. When n is a positive integer:

$$(a+b)^n = a^n + na^{n-1}b + \frac{n(n-1)}{2!}a^{n-2}b^2 + \frac{n(n-1)(n-2)}{3!}a^{n-3}b^3 + \ldots + b^n.$$

The situation where n is *not* a positive integer is dealt with in Core 4.

This expansion may appear complicated but it starts to look simpler if you follow the patterns from one term to the next:

Remember that $a^0 = 1$ when $a \neq 0$.

- Starting with a^n, the power of a is reduced by 1 each time until the last term, b^n, which is the same as $a^0 b^n$.

- The power of b is increased by 1 each time. Notice that the first term, a^n, is the same as $a^n b^0$ and that the powers of a and b always add up to n in each term.

- It's worth remembering the first three coefficients: 1, n and $\dfrac{n(n-1)}{2!}$ then the rest follow the same pattern, giving $\dfrac{n(n-1)(n-2)}{3!}$, $\dfrac{n(n-1)(n-2)(n-3)}{4!}$ and so on.

Pascal's triangle: each row starts and ends with 1. Every other value is found by adding the pair of numbers immediately above it in the pattern.

```
      1    4    6
  1   5   10
```

The coefficients in the expansion of $(a+b)^n$ also follow the pattern given by the row of **Pascal's triangle** that starts with 1 n …
So the coefficients in the expansion of $(a+b)^4$ for example are 1, 4, 6, 4 and 1.

```
                1
             1     1
          1     2     1
       1     3     3     1
    1     4     6     4     1
 1     5    10    10     5     1
```

It follows that
$(a+b)^4 = a^4 + 4a^3b + 6a^2b^2 + 4ab^3 + b^4.$

Variations on this result can be obtained by substituting different values for a and b. Some examples are:

In this case, $a=1$ and $b=x$.

$$(1+x)^4 = 1 + 4x + 6x^2 + 4x^3 + x^4.$$

and

For this one, $a=1$ and $b=-2x$.

$$(1-2x)^4 = 1 + 4(-2x) + 6(-2x)^2 + 4(-2x)^3 + (-2x)^4$$
$$= 1 - 8x + 24x^2 - 32x^3 + 16x^4.$$

The notation $\dbinom{n}{r}$ is often used to stand for the expression $\dfrac{n(n-1)\ldots(n-r+1)}{r!}$.

So, for example $\dbinom{n}{3} = \dfrac{n(n-1)(n-2)}{3!}.$

Using this notation, the binomial expansion may be written as:

$$(a+b)^n = a^n + \binom{n}{1}a^{n-1}b + \binom{n}{2}a^{n-2}b^2 + \binom{n}{3}a^{n-3}b^3 + \ldots + b^n.$$

- It's useful to recognise that the term involving b^r takes the form $\binom{n}{r}a^{n-r}b^r$.

- For positive integer values $\binom{n}{r}$ has the same value as ^nC_r, and you may find that your calculator will work this out for you.

- When n is small and the full binomial expansion is required, the simplest way to find the coefficients is to use Pascal's triangle. For larger values of n it may be simpler to use the formula, particularly if only some of the terms are required.

Example Expand $(1 + 3x)^{10}$ in ascending powers of x up to and including the fourth term.

$$(1+3x)^{10} = 1 + \binom{10}{1}(3x) + \binom{10}{2}(3x)^2 + \binom{10}{3}(3x)^3 + \ldots$$

$$= 1 + 10(3x) + 45(9x^2) + 120(27x^3) + \ldots$$

$$= 1 + 30x + 405x^2 + 3240x^3 + \ldots.$$

Example Find the coefficient of the x^7 term in the expansion of $(3 - 2x)^{15}$

Notice that the coefficient of x^7 is not totally determined by the value of $\binom{15}{7}$.

The term involving x^7 is given by $\binom{15}{7}(3)^8(-2x)^7$

$$= 6435 \times 6561 \times -128x^7,$$

and so the required coefficient is $-5\,404\,164\,480$.

Progress check

1 Find the sum of the first 20 terms of $12 + 15 + 18.75 + \ldots$ to 4 s.f.

2 Explain why the series $10 + 9 + 8.1 + 7.29 + \ldots$ is convergent and find the value of its sum to infinity to 5 s.f.

3 Expand $(1 + 3x)^{12}$ up to and including the third term.

3 $1 + 36x + 594x^2$
2 $|r| = 0.9 < 1, 100.$
1 4115

2.4 Trigonometry

After studying this section you should be able to:

- state the exact values of the sine, cosine and tangent of special angles
- understand the properties of the sine, cosine and tangent functions
- sketch the graphs of trigonometric functions
- use the sine and cosine rules and the formula for the area of a triangle
- understand radian measure and use it to find arc lengths and areas of sectors
- solve trigonometric equations in a given interval

LEARNING SUMMARY

Sine, cosine and tangent of special angles

The sine, cosine and tangent of 30°, 45°, and 60° may be expressed exactly as shown below. The results are found from Pythagoras' theorem.

$$\sin 30° = \frac{1}{2} \qquad \sin 60° = \frac{\sqrt{3}}{2} \qquad \sin 45° = \frac{1}{\sqrt{2}}$$

$$\cos 30° = \frac{\sqrt{3}}{2} \qquad \cos 60° = \frac{1}{2} \qquad \cos 45° = \frac{1}{\sqrt{2}}$$

$$\tan 30° = \frac{1}{\sqrt{3}} \qquad \tan 60° = \sqrt{3} \qquad \tan 45° = 1.$$

Trigonometric graphs

You need to be able to sketch the graphs of the sine, cosine and tangent functions and to know their special properties.

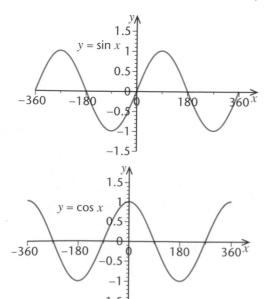

$\sin x$ is defined for any angle and always has a value between −1 and 1. It is a **periodic function** with period 360°.

The graph has **rotational symmetry** of order 2 about every point where it crosses the x-axis.

It has **line symmetry** about every vertical line passing through a vertex.

$\cos x \equiv \sin(x + 90°)$ so the graph of $y = \cos x$ can be obtained by translating the sine graph 90° to the left.

It follows that $\cos x$ is also a periodic function with period 360° and has the corresponding symmetry properties.

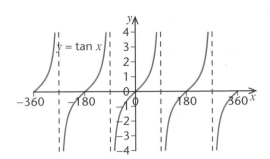

$$\tan x \equiv \frac{\sin x}{\cos x}.$$

$\tan x$ is undefined whenever $\cos x = 0$ and approaches $\pm\infty$ near these values. It is a periodic function with period $180°$.

The graph has rotational symmetry of order 2 about $0°$, $\pm 90°$, $\pm 180°$, $\pm 270°$, ...

The graphs of more complex trigonometric functions can often be produced by applying transformations to one of the basic graphs (see Core 1, page 18–19).

Example Sketch the graph of $y = 2\sin(x + 30°) + 1$ for $0° \leqslant x \leqslant 360°$.

It is helpful to think about building the transformations in stages.

Transformations			
Basic function	Translate the curve 30° to the left.	Now apply a one-way stretch with scale factor 2 parallel to the y-axis.	Finally translate the curve 1 unit up.
$y = \sin x$	$y = \sin(x + 30)$	$y = 2\sin(x + 30)$	$y = 2\sin(x + 30°) + 1$

The graph of $y = 2\sin(x + 30°) + 1$ may now be produced by applying these transformations in order starting from the graph of $y = \sin x$.

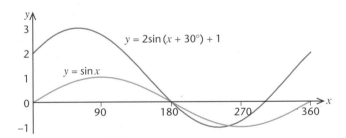

Area of a triangle

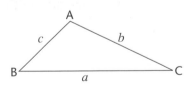

The same formula applies whether the angle is acute or obtuse.

Area of triangle $= \frac{1}{2} ab \sin C$

In both the diagrams, $b \sin C$ represents the height of the triangle.

The sine and cosine rules

A, B and C can be at any vertex. The opposite sides are then labelled as a, b, c.

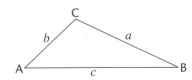

Sine rule: $\dfrac{a}{\sin A} = \dfrac{b}{\sin B} = \dfrac{c}{\sin C}$

Cosine rule: $a^2 = b^2 + c^2 - 2bc \cos A$

This may be rearranged to find an angle

$$\cos A = \frac{b^2 + c^2 - a^2}{2bc}$$

Radians

Angles can also be measured in **radians** and this makes it much easier to deal with trigonometric functions when using **calculus**.

1 radian $\approx 57.3°$. This may be written as $1^c \approx 57.3°$. However, the symbol for radians is not normally written when the angle involves π.

The following results are useful to remember:

$$\pi = 180°, \frac{\pi}{2} = 90°, \frac{\pi}{4} = 45°, \frac{\pi}{3} = 60°, \frac{\pi}{6} = 30°.$$

All the results that you know in degrees also apply in radians, e.g. $\sin\frac{\pi}{6} = 0.5$

Arc length and sector area

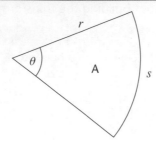

With θ in radians:

- Length of arc of sector is given by
  $s = r\theta$.

- Area of sector is given by
  $A = \frac{1}{2}r^2\theta$.

Solving trigonometric equations

You may be required to solve trigonometric equations in a variety of forms.

Example Solve $\sin(2x - 30°) = 0.6$ for $0° \leqslant x \leqslant 360°$.

> This is the principal value given by the calculator.

First find a value of $2x - 30°$.

One value of $2x - 30°$ is $\sin^{-1}(0.6) = 36.86...°$.
You need to adjust the interval to find where the values of $2x - 30°$ must lie.

$$0° \leqslant x \leqslant 360° \Rightarrow 0° \leqslant 2x \leqslant 720° \Rightarrow -30° \leqslant 2x - 30° \leqslant 690°.$$

Now look at the graph of the sine function in this interval.

This is the graph of $y = \sin X$ for $-30° \leqslant X \leqslant 690°$ where $X = 2x - 30°$.
In other words, draw the sine curve as normal for the interval and then interpret the variable as $2x - 30°$.

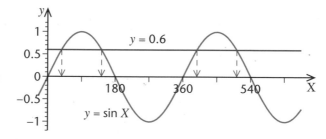

The graph shows that there are four values of $2x - 30°$ in the interval that make the equation work.

Using the symmetry of the curve, the values of $2x - 30°$ are:

$36.86...°, 180° - 36.86...°, 360° + 36.86...°, 540° - 36.86...°.$

This gives:

$2x - 30° = 36.86...° \Rightarrow x = 33.4°$

$2x - 30° = 143.14...° \Rightarrow x = 86.6°$

$2x - 30° = 396.86...° \Rightarrow x = 213.4°$

$2x - 30° = 503.14...° \Rightarrow x = 266.6°$

So $x = 33.4°, 86.6°, 213.4°, 266.6°$ (1 d.p.).

Using trigonometric identities

The following trigonometric relationships are **identities**. This means that they are true for all values of x.

They are very important when simplifying expressions and solving equations and should be learnt.

> $$\tan x = \frac{\sin x}{\cos x}$$
>
> KEY POINT

Example
Solve the equation $\sin x = \sqrt{3}\,\cos x$, for values of x between $0°$ and $360°$.

This is allowed since $\cos x = 0$ is not a solution.

Divide each side by $\cos x$.

$$\frac{\sin x}{\cos x} = \sqrt{3}$$

$$\Rightarrow \tan x = \sqrt{3}$$

$$x = 60° \text{ or } 240°$$

The graph of $y = \tan x$ repeats every $180°$, so to find the second solution, add $180°$

> $$\sin^2 x + \cos^2 x = 1$$
>
> KEY POINT

This is sometimes known as the Pythagorean trigonometric identity.

Example Solve the equation $\sin^2 x - \cos^2 x - \sin x = 1$ for $-\pi \leqslant x \leqslant \pi$.

*You need to recognise that this is a quadratic in $\sin x$.
It has to be rearranged in order to use the quadratic formula.*

Replacing $\cos^2 x$ with $1 - \sin^2 x$ gives $\sin^2 x - (1 - \sin^2 x) - \sin x = 1$

$$\Rightarrow 2\sin^2 x - \sin x - 2 = 0.$$

$-1 \leqslant \sin x \leqslant 1$ for all values of x.

Using the formula gives $\sin x = \dfrac{1 \pm \sqrt{1 + 16}}{4} = \dfrac{1 \pm \sqrt{17}}{4}$

$$\Rightarrow \sin x = 1.28 \ldots \text{ (no solutions) or } \sin x = -0.78077 \ldots.$$

Remember to set your calculator to radians mode.

Using $\sin^{-1}$ gives $\qquad x = -0.8959^c$ to 4 d.p.

Another solution in the interval
$-\pi \leqslant x \leqslant \pi$ is $-\pi + 0.8959^c$ giving
$x = -2.2457^c$ to 4 d.p.

The diagram shows there are no other solutions in this interval.

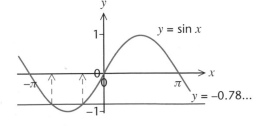

Progress check

1 Find the exact value of:

(a) $\sin 120°$ (b) $\cos 135°$ (c) $\tan 240°$.

2 Describe the transformations needed to turn the graph of $y = \cos x$ into the graph of:

(a) $y = \cos 2x$ (b) $y = 3 \cos x$ (c) $y = 3 \cos(2x) - 1$.

3 Find the value of x and θ.

4 Find the exact value of:

(a) $\sin \dfrac{\pi}{4}$ (b) $\cos \dfrac{\pi}{3}$ (c) $\tan \dfrac{\pi}{6}$.

5 Solve $8 \sin x = 3 \cos x$ for $-360° \leqslant x \leqslant 360°$.

6 Solve $\cos(3x + 20°) = 0.6$ for $0° \leqslant x \leqslant 180°$.

7 Solve $2 \sin^2 x + \cos x = 1$ for $-\pi \leqslant x \leqslant \pi$.

8 OAB is a sector of a circle, centre O, radius 4 cm.
Angle AOB = 0.5 radians.

(a) Find the perimeter of sector AOB.

(b) Find the length of the chord AB.

(c) Find the area of the sector OAB.

(d) Find the area of the triangle OAB.

(e) Find the area of the shaded segment.

8 (a) 10 cm (b) 1.98 cm (2 d.p.) (c) 4 cm² (d) 3.84 cm² (2 d.p.) (e) 0.16 cm² (2 d.p.)

7 $\dfrac{2\pi}{3}$, 0, $-\dfrac{2\pi}{3}$

6 11.0°, 95.6°, 131.1°

5 $\tan x = \frac{3}{8}$ giving $x = 200.6°$, 20.6°, $-159.4°$ and $-339.4°$

4 (a) $\dfrac{1}{\sqrt{2}}$ (b) $\frac{1}{2}$ (c) $\dfrac{1}{\sqrt{3}}$

3 $x = 12.0$ m (3 s.f.) $\theta = 36.6°$ (1 d.p.).

2 (a) One-way stretch with scale factor 0.5 parallel to the x-axis.
(b) One-way stretch with scale factor 3 parallel to the y-axis.
(c) One-way stretch with scale factor 0.5 parallel to the x-axis, followed by a one-way stretch with scale factor 3 parallel to the y-axis, followed by a translation of 1 unit downwards.

1 (a) $\dfrac{\sqrt{3}}{2}$ (b) $-\dfrac{1}{\sqrt{2}}$ (c) $\sqrt{3}$

2.5 Exponentials and logarithms

After studying this section you should be able to:

- understand exponential functions and sketch their graphs
- understand the log laws and use them to solve equations of the form $a^x = b$

LEARNING SUMMARY

Exponential functions

An **exponential function** is one where the variable is a power or exponent. For example, any function of the form $f(x) = a^x$, where a is a constant, is an exponential function.

The diagram shows the graphs of $y = 2^x$ and $y = 10^x$.

Exponential functions are often used to represent, or model, patterns of:
- growth when $a > 1$
- decay when $a < 1$.

All graphs of the form $y = a^x$ cross the y-axis at (0, 1). Each graph has a different gradient at this point and the value of the gradient depends on a.

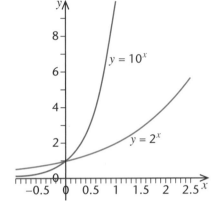

Logarithms

Exponential functions are related to **logarithmic functions** as follows:

$$y = \log_a x \Leftrightarrow x = a^y$$

$a > 0$ and $a \neq 1$.

a is called the base of the logarithm.

The **logarithm** (log) to a given **base** of a number is the **power** to which the base must be raised to equal the number.

Your calculator has a key for logs to the base 10, probably labelled log.

For example, $10^2 = 100$, so $\log_{10} 100 = 2$.

It is useful to remember the following hold for any base a:

$\log_a a = 1$, since $a^1 = a$

$\log_a 1 = 0$, since $a^0 = 1$.

The graph of $y = \log_a x$ is a reflection of $y = a^x$ in the line $y = x$.

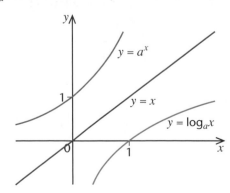

The log laws

There are three laws of logarithms that you need to know. The same laws apply in any base and so no particular base is stated:

- $\log a + \log b = \log ab$

- $\log a - \log b = \log\left(\dfrac{a}{b}\right)$

- $n \log a = \log a^n$.

You can use these laws to simplify expressions involving logs and to solve **exponential equations**, i.e. equations where the unknown value is a power.

Example Express $\log x + 3 \log y$ as a single logarithm.

$$\log x + 3 \log y = \log x + \log y^3 \qquad \text{(using the third law)}$$
$$= \log xy^3 \qquad \text{(using the first law)}$$

Example Solve the equation $5^x = 30$.

Taking logarithms of both sides gives:

$$\log 5^x = \log 30$$
$$\Rightarrow x \log 5 = \log 30$$
$$\Rightarrow x = \frac{\log 30}{\log 5} = 2.113 \text{ to 4 s.f}$$

This works in any base.

Change of the base of the logarithm

You can use the change of base formula to work out logarithms using any base.

$$\log_a b = \frac{\log_c b}{\log_c a}$$

$$\text{So } \log_7 6 = \frac{\log_{10} 6}{\log_{10} 7} = 0.9208$$

Progress check

1 Solve the equation $4^x = 100$ and give your answer to 4 s.f.

2 Write each of the following as a single logarithm.
(a) $\log_a x + 3 \log_a y - \frac{1}{2} \log_a z$
(b) $\log_{10} x - 1$

3 Evaluate the following:
(a) $\log_2 8$
(b) $\log_9 3$
(c) $2 \log_4 2$

3 (a) 3 (b) $\frac{1}{2}$ (c) 1

2 (a) $\log_a\left(\dfrac{xy^3}{\sqrt{z}}\right)$ (b) $\log_{10}\left(\dfrac{x}{10}\right)$

1 $x = 3.322$

2.6 Differentiation

After studying this section you should be able to:

- recall differentiation methods and applications required in Core 1
- locate stationary points and use the second derivative method to distinguish between maxima and minima
- determine whether a function is increasing or decreasing in an interval

LEARNING SUMMARY

Curve sketching

The gradient function gives information about the behaviour of a curve.

$f'(x) > 0$ at A and the function is **increasing**.

$f'(x) < 0$ at C and the function is **decreasing**.

$f'(x) = 0$ at B and D and the function is neither increasing nor decreasing. These points are called **stationary points**. B is at a **local maximum** and D is at a **local minimum**.

> These types of stationary point are also known as **turning points**.

E and F show a different type of stationary point, called a **stationary point of inflexion**.

One way to distinguish between the types of stationary point is to look at the sign of the derivative on either side of the point.

- At a local maximum the sign of the derivative changes from positive to negative.
- At a local minimum the sign of the derivative changes from negative to positive.
- At a stationary point of inflexion there is no change of sign.

The second derivative

Starting from $\dfrac{dy}{dx} = f'(x)$ and differentiating again gives the **second derivative**. This is written as $\dfrac{d^2y}{dx^2} = f''(x)$. The second derivative can give information about the nature of any stationary points.

At a stationary point:
- $f''(x) > 0 \Rightarrow$ the point is a local minimum
- $f''(x) < 0 \Rightarrow$ the point is a local maximum.

But, if $f''(x) = 0$ then this gives no further information.

Example   Find the stationary points of the curve $y = \dfrac{x^3}{3} - \dfrac{3}{2}x^2 + 2x - 1$ and use the second derivative to distinguish between them.

First differentiate the equation, then solve $\dfrac{dy}{dx} = 0$ to locate the stationary points.

$$\frac{dy}{dx} = x^2 - 3x + 2.$$

At a stationary point $\dfrac{dy}{dx} = 0 \Rightarrow x^2 - 3x + 2 = 0$

$$\Rightarrow (x-1)(x-2) = 0$$
$$\Rightarrow x = 1 \text{ or } x = 2.$$

When $x = 1$, $y = \frac{1}{3} - \frac{3}{2} + 2 - 1 = -\frac{1}{6}$.

When $x = 2$, $y = \frac{8}{3} - 6 + 4 - 1 = -\frac{1}{3}$.

The stationary points are $(1, -\frac{1}{6})$ and $(2, -\frac{1}{3})$.

In this case, the function is a cubic and so the shape of the curve is known. We should expect the first stationary point to be a local maximum and the second to be a local minimum.

Using the second derivative: $f''(x) = 2x - 3$

so $f''(1) = -1 < 0$ giving a local maximum,

and $f''(2) = 1 > 0$ giving a local minimum as expected.

Example This sheet of metal is bent to form an open box. Find the depth of the box which gives a maximum volume.

$V = x(30 - 2x)(30 - 2x) = 900x - 120x^2 + 4x^3$

$\dfrac{dV}{dx} = 900 - 240x + 12x^2 = 12(15 - x)(5 - x)$

$\dfrac{dV}{dx} = 0$ at turning points $x = 15$ and $x = 5$

$\dfrac{d^2V}{dx^2} = -240 + 24x$

$x = 15 \quad \dfrac{d^2V}{dx^2} = 120$ minimum

$x = 5 \quad \dfrac{d^2V}{dx^2} = -120$ maximum

Maximum volume at depth 5cm.

Progress check

1 Locate the stationary points on the curve $y = 2x^3 - 6x^2 - 18x + 5$ and use the second derivative to distinguish between them.

2 A rectangle of length x, where x varies, has a constant area of 48cm². Express the perimeter, y, in terms of x. Find the least possible value of y.

Solve the inequality $f'(x) < 0$.

3 Find the values of x for which the function $f(x) = x^3 - 6x^2 + 9x - 7$ is decreasing.

3 $1 < x < 3$

2 $y = 16\sqrt{3}$cm

1 Local maximum: $(-1, 15)$; local minimum $(3, -49)$.

2.7 Integration

After studying this section you should be able to:

- evaluate a definite integral
- find the area under a curve
- use the trapezium rule to find an approximate value of an integral

LEARNING SUMMARY

Indefinite integration

Indefinite integration was covered in Core 1; for revision, you should look at pages 31 to 32 and tackle the Progress check at the end of the section.

Example 1

$$\int (x^{\frac{3}{2}} + 2x^{-\frac{1}{2}})dx = \frac{x^{\frac{5}{2}}}{\frac{5}{2}} + 2\frac{x^{\frac{1}{2}}}{\frac{1}{2}} + c$$

$$= \frac{2}{5}x^{\frac{5}{2}} + 4x^{\frac{1}{2}} + c$$

Example 2

Write the expression in index form first.

$$\int \frac{x+2}{\sqrt{x}}\ dx = \int (x^{\frac{1}{2}} + 2x^{-\frac{1}{2}})dx$$

$$= \frac{2}{3}x^{\frac{3}{2}} + 4x^{\frac{1}{2}} + c$$

Definite integration

An integral of the form $A = \int_a^b f(x)dx$ is a **definite integral**.

It has a numerical value and is evaluated as follows:

Substitute the upper limit, then subtract the value when you substitute the lower limit.

$$\int_a^b f(x)dx = [g(x)]_a^b = g(b) - g(a)$$

KEY POINT

A constant of integration is **not** needed in a definite integral.

For example $\int_1^4 2x\ dx = [x^2]_1^4 = 4^2 - 1^2 = 15.$

Area under a curve

For areas *below* the x-axis, the definite integral gives a *negative* value.

The area enclosed by the curve $y = f(x)$, the x-axis and the lines $x = a$ and $x = b$ is given by $\int_a^b ydx.$

KEY POINT

Example Find the area under the curve $y = x^2 + 1$ between $x = 1$ and $x = 3$.

53

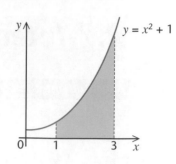

$$\text{Area} = \int_1^3 (x^2 + 1)\, dx = \left[\frac{x^3}{3} + x\right]_1^3$$

$$= \left(\frac{3^3}{3} + 3\right) - \left(\frac{1^3}{3} + 1\right)$$

$$= 9 + 3 - \frac{1}{3} - 1$$

$$= 10\tfrac{2}{3}.$$

To find the **area between a line and a curve**, find the area under the line and the area under the curve separately and then subtract to find the required area. A sketch will be useful.

Progress check

1 Find the value of each of these definite integrals.

(a) $\displaystyle\int_0^5 (x - 2)\, dx$ (b) $\displaystyle\int_4^9 \sqrt{x}\, dx$ (c) $\displaystyle\int_0^6 y^2\, dy$

2 Find the area under the curve $y = 3\sqrt{x}$
between $x = 1$ and $x = 9$.

2 52

1 (a) 2.5 (b) 12⅔ (c) 72

Numerical integration – trapezium rule

The value of $A = \displaystyle\int_a^b y\,dx$ represents the area under the graph of $y = f(x)$ between $x = a$ and $x = b$. You can find an approximation to this value using the **trapezium rule**. This is especially useful if the function is difficult to integrate.

The area under the curve between a and b may be divided into n strips of equal width h.

It follows that $h = \dfrac{b - a}{n}$.

Each strip is approximately a trapezium and so the total area is approximately

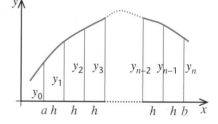

$$\tfrac{1}{2}h(y_0 + y_1) + \tfrac{1}{2}h(y_1 + y_2) + \tfrac{1}{2}h(y_2 + y_3) + \ldots + \tfrac{1}{2}h(y_{n-2} + y_{n-1}) + \tfrac{1}{2}h(y_{n-1} + y_n).$$

> To increase the accuracy, use more strips.

KEY POINT

This simplifies to give the formula known as the **trapezium** rule:

$$\int_a^b y\,dx = \tfrac{1}{2}h\{(y_0 + y_n) + 2(y_1 + y_2 + \ldots + y_{n-1})\} \text{ where } h = \frac{b - a}{n}.$$

Example

Use the trapezium rule with five strips to estimate the value of $\displaystyle\int_1^2 2^x\,dx$.

It's useful to tabulate the information:

> Work to a greater level of accuracy in your table than you intend to give in your final answer.

x	1	1.2	1.4	1.6	1.8	2
2^x	2	2.2974	2.6390	3.0314	3.4822	4
	y_0	y_1	y_2	y_3	y_4	y_5

$$h = \frac{2-1}{5} = 0.2$$

So, $\displaystyle\int_1^2 2^x \, dx \approx 0.1\{(2+4) + 2(2.2974 + 2.6390 + 3.0314 + 3.4822)\}$

$$= 2.89 \quad \text{to} \quad 3 \text{ s.f.}$$

Increasing gradient: overestimate

Decreasing gradient: underestimate

In this case, the working was done to 5 s.f. and the final answer is given to 3 s.f.

If the gradient of the graph is increasing over the interval then the trapezium rule will give an over-estimate of the area. If the gradient of the graph is decreasing over the interval then it will give an under-estimate of the area.

A graphic calculator with a numerical integration function gives the value of the area as 2.8853901

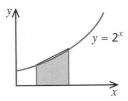

In this case, the gradient is increasing over the interval and so the trapezium rule gives an over-estimate of the area.

Progress check

1 Use the trapezium rule with five strips to estimate $\displaystyle\int_1^2 \frac{1}{x} \, dx$.

State whether the value obtained is an over-estimate or an under-estimate.

1 0.69563 (5 d.p.)
Over-estimate

Sample questions and model answers

1

Find the exact values of x, where $0 \leqslant x \leqslant 2\pi$, that satisfy the equation

$$\tan x = 2 \sin x.$$

Use the identity
$\tan x = \dfrac{\sin x}{\cos x}$

Multiply both sides by $\cos x$.

Do not cancel $\sin x$ as this leads to loss of solutions.

These are exact values.

$$\tan x = 2 \sin x$$

$$\frac{\sin x}{\cos x} = 2 \sin x$$

$$\sin x = 2 \sin x \cos x$$

$$\sin x - 2 \sin x \cos x = 0$$

$$\sin x(1 - 2 \cos x) = 0$$

Either $\quad \sin x = 0 \Longrightarrow x = 0, \pi, 2\pi$

or $\quad 1 - 2 \cos x = 0 \Longrightarrow x = \dfrac{\pi}{3}, \dfrac{5\pi}{3}$

So $x = 0, \dfrac{\pi}{3}, \pi, \dfrac{5\pi}{3}, 2\pi$

2

(a) Given that $\log_a x - \log_a 8 + 2 \log_a 4 = 0$, where a is a positive constant, find x.

(b) Find the value of x, correct to two decimal places, that satisfies the equation
$3^{2x+1} = 40$.

Simplify, using the log laws.

Remember that $a^0 = 1$ for all values of a.

Take logs to the base 10 of both sides.

Use $\log_a(b^c) = c\log_a b$

(a) $$\log_a x - \log_a 8 + 2\log_a 4 = 0$$

$$\Longrightarrow \quad \log_a x - \log_a 16 + \log_a 8 = 0$$

$$\Longrightarrow \quad \log_a\left(\frac{16x}{8}\right) = 0$$

$$\Longrightarrow \quad \log_a(2x) = 0$$

$$\Longrightarrow \quad 2x = 1$$

$$x = 0.5$$

(b) $$3^{2x+1} = 40$$

$$\log_{10}(3^{2x+1}) = \log_{10} 40$$

$$(2x+1)\log_{10} 3 \doteqdot \log_{10} 40$$

$$2x+1 = \frac{\log_{10} 40}{\log_{10} 3} = 3.3577\ldots$$

$$2x = 2.3577\ldots$$

$$x = 1.18 \; (2 \text{ d.p.})$$

Sample questions and model answers *(continued)*

3

The third term of a geometric series is 20 and the fifth term is 5.

Given that the common ratio, r, is negative, find

(a) the value of r

(b) the first term,

(c) the sum to infinity of the series.

The nth term of a geometric series is ar^{n-1}.

(a) $ar^2 = 20$ (1)

$ar^4 = 5$ (2)

Dividing (2) by (1) gives $r^2 = \frac{1}{4}$

$r < 0 \Rightarrow r = -\frac{1}{2}$.

(b) Substitute into (1) $a(\frac{1}{4}) = 20$

$a = 80$

The series is convergent, since $|r| < 1$.

(c) $s_\infty = \dfrac{a}{1-r}$

$= \dfrac{80}{1-(-\frac{1}{2})}$

$= 53\frac{1}{3}$

4

(a) Express $\dfrac{5x^2 - 1}{\sqrt{x}}$ in the form $5x^a - x^b$, where a and b are rational numbers to be found.

(b) Hence find the exact value of $\displaystyle\int_1^2 \dfrac{5x^2 - 1}{\sqrt{x}}\ dx$.

(a) $\dfrac{5x^2 - 1}{\sqrt{x}} = \dfrac{5x^2}{x^{1/2}} - \dfrac{1}{x^{1/2}} = 5x^{3/2} - x^{-1/2}$

Remember that $\sqrt{x} = x^{1/2}$ and work in index form.

So $a = \frac{3}{2}$ and $b = -\frac{1}{2}$.

(b) $\displaystyle\int_1^2 \dfrac{5x^2 - 1}{\sqrt{x}}\ dx = \int_1^2 (5x^{3/2} - x^{-1/2})\ dx$

$= \left[\dfrac{5x^{5/2}}{\frac{5}{2}} - \dfrac{x^{1/2}}{\frac{1}{2}} \right]_1^2$

$= [2x^{5/2} - 2x^{1/2}]_1^2$

$(2)^{5/2} = (\sqrt{2})^5 = 4\sqrt{2}$

$= (8\sqrt{2} - 2\sqrt{2}) - (2 - 2)$

Leave the answer in surd form.

$= 6\sqrt{2}$

Practice examination questions

1 (a) Find the 8th term of a geometric progression with first term 3 and common ratio 2.

(b) Find the sum of the first 50 terms of the series $25 + 26.2 + 27.4 + \ldots$.

2 (a) Given that $(x + 2)$ is a factor of $f(x) = x^3 - 4x^2 - 3x + k$, use the factor theorem to find the value of k.

(b) Factorise $f(x)$ completely and sketch the graph of $y = f(x)$.

(c) Solve the inequality $f(x) \geqslant 0$.

(d) Find the remainder when $f(x)$ is divided by $(x + 1)$.

3 (a) Express $3 \log_a x - 2 \log_a y + \log_a (x + 1)$ as a single logarithm.

(b) Solve the equation $5^x = 100$. Give your answer to 3 d.p.

4 Solve the equation $\cos(2x + 30°) = 0.4$ for $0° \leqslant x \leqslant 360°$.

5 (a) Write the equation $\cos x + 3 \sin x \tan x - 2 = 0$ in the form $a \cos^2 x + b \cos x + c = 0$.

(b) Find the solutions of $\cos x + 3 \sin x \tan x - 2 = 0$ for $0° \leqslant x \leqslant 360°$.

6 Given that $f(x) = x^2 + \dfrac{1}{x}$, find

(a) $f'(2)$.

(b) $f''(-1)$.

7 The diagram shows a region R bounded by the curve $y = 3\sqrt{x}$, the x-axis and the lines $x = 1$ and $x = 4$.

Calculate the area of the region R.

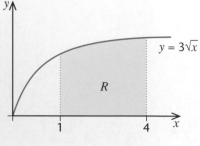

8 The diagram shows a circle of unit radius.

The angle between the radii is θ where

$0 < \theta < \pi$ and the area of the shaded

segment is $\dfrac{\pi}{5}$.

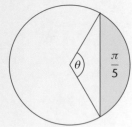

Show that $\theta = \sin \theta + \dfrac{2\pi}{5}$.

Practice examination questions (continued)

9 The area shaded in the diagram is bounded by the curve $y = \sqrt{x^2 - 3}$, the lines $x = 2$ and $x = 3$ and the x-axis.

Express the shaded area as an integral and estimate its value using the trapezium rule with five intervals. Give your answer to 2 d.p.

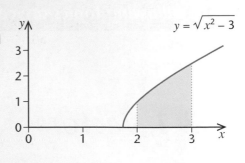

10 (a) Factorise $2x^{\frac{3}{2}} - \dfrac{2x^{\frac{5}{2}}}{5}$.

(b) Find the points at which the graph of $y = 2x^{\frac{3}{2}} - \dfrac{2x^{\frac{5}{2}}}{5}$ meets the x-axis.

(c) Find $\dfrac{dy}{dx}$.

(d) Solve $\dfrac{dy}{dx} = 0$ for $x > 0$. Hence find the coordinates of the turning point of the graph, where x is positive, and determine its nature.

11 (a) Find the coordinates of the stationary points on the curve

$$y = 2x^3 - 15x^2 - 36x + 10.$$

(b) Use the second derivative to determine the nature of the stationary points found in (a).

(c) Find the set of values of x for which $2x^3 - 15x^2 - 36x + 10$ is a decreasing function of x.

Chapter 3
Mechanics 1

The following topics are covered in this chapter:

- *Vectors*
- *Kinematics*

- *Statics and moments*
- *Dynamics*

3.1 Vectors

After studying this section you should be able to:

- *understand the distinction between vector and scalar quantities*
- *add and subtract vector quantities and multiply by a scalar*
- *resolve vector quantities into two perpendicular components*
- *use the unit vectors $\mathbf{i}$, $\mathbf{j}$ and $\mathbf{k}$*
- *find the resultant, magnitude and direction of a vector*
- *understand the application of vectors in mechanics*

LEARNING SUMMARY

Vector and scalar quantities

A **scalar** quantity has size (or **magnitude**) but not direction. **Numbers** are scalars and some other important examples are **distance**, **speed**, **mass** and **time**.

A **vector** quantity has both size and **direction**. For example, distance in a specified direction is called **displacement**. Some other important examples are **velocity**, **acceleration**, **force** and **momentum**.

The diagram shows a **directed line segment**. It has size (in this case, length) and direction so it is a vector.

The diagram gives a useful way to represent *any* vector quantity and may also be used to represent addition and subtraction of vectors and multiplication of a vector by a scalar.

In a textbook, vectors are usually labelled with lower case letters in **bold** print.
When hand-written, these letters should be underlined e.g. $\underline{a}$

Addition and subtraction of vectors

This diagram shows three vectors $\mathbf{a}$, $\mathbf{b}$ and $\mathbf{c}$ such that $\mathbf{c} = \mathbf{a} + \mathbf{b}$.

The vectors $\mathbf{a}$ and $\mathbf{b}$ follow on from each other and then $\mathbf{c}$ joins the start of vector $\mathbf{a}$ to the end of vector $\mathbf{b}$.

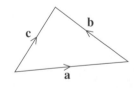

Check that $\mathbf{a} + \mathbf{b} = \mathbf{b} + \mathbf{a}$.

The vector $\mathbf{c}$ is called the **resultant** of $\mathbf{a}$ and $\mathbf{b}$.

On a vector diagram, $-\mathbf{q}$ has the opposite sense of direction to $\mathbf{q}$.
Notice that $\mathbf{p} - \mathbf{q}$ is represented as $\mathbf{p} + (-\mathbf{q})$.

This diagram shows the same information in a different way.

Following the route in the opposite direction to $\mathbf{q}$ is the same as adding $-\mathbf{q}$ or subtracting $\mathbf{q}$.

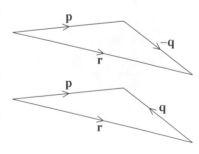

Scalar multiplication

2**p** is parallel to **p** and has the same sense of direction but is twice as long.

−3**p** is parallel to **p** but has the opposite sense of direction and is three times as long.

Component form

When working in two dimensions, it is often very useful to express a vector in terms of two special vectors **i** and **j**. These are **unit vectors** at right-angles to each other. A vector **r** written as **r** = a**i** + b**j** is said to have **components** a**i** and b**j**.

> A unit vector is a vector of magnitude 1 unit.

Examples

r = **i** + 2**j**

r = 3**i** − **j**

For work involving the Cartesian coordinate system, **i** and **j** are taken to be in the positive directions of the x- and y-axes respectively.

In three dimensions, a third vector **k** is used to represent a unit vector in the positive direction of the z-axis.

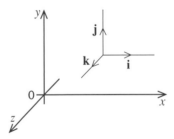

The **position vector** of a point P is the vector $\overrightarrow{OP}$ where O is the origin.

If P has coordinates (a, b) then its position vector is given by **r** = a**i** + b**j**.

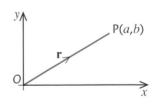

In three dimensions, a point with coordinates (a, b, c) would have position vector **r** = a**i** + b**j** + c**k**.

Resolving a vector

If you know the magnitude and direction of a vector then you can use trigonometry to **resolve** it in terms of **i** and **j** components.

The magnitude of **r** may be written as $|\mathbf{r}|$ or simply as r.

From the diagram, **r** = $r \cos \theta$**i** + $r \sin \theta$**j**.

Example

A force **f** has magnitude 10 N at an angle of 60°
above the horizontal. Express **f** in the form $a\mathbf{i} + b\mathbf{j}$
where **i** and **j** are unit vectors in the horizontal and
vertical directions respectively.

In this case, $a = 10 \cos 60° = 5$ and
$b = 10 \sin 60° = 5\sqrt{3}$ giving $\mathbf{f} = 5\mathbf{i} + 5\sqrt{3}\mathbf{j}$.

Adding and subtracting vectors in component form

One advantage of expressing vectors in terms of **i** and **j** is that addition and
subtraction may be done algebraically.

Example

$\mathbf{p} = 3\mathbf{i} + 2\mathbf{j}$ and $\mathbf{q} = 5\mathbf{i} - \mathbf{j}$. Express the following vectors in terms of **i** and **j**:

(a) $\mathbf{p} + \mathbf{q}$ (b) $\mathbf{p} - \mathbf{q}$ (c) $2\mathbf{p} - 3\mathbf{q}$.

(a) $\mathbf{p} + \mathbf{q} = (3\mathbf{i} + 2\mathbf{j}) + (5\mathbf{i} - \mathbf{j}) = 8\mathbf{i} + \mathbf{j}$

(b) $\mathbf{p} - \mathbf{q} = (3\mathbf{i} + 2\mathbf{j}) - (5\mathbf{i} - \mathbf{j}) = -2\mathbf{i} + 3\mathbf{j}$

(c) $2\mathbf{p} - 3\mathbf{q} = 2(3\mathbf{i} + 2\mathbf{j}) - 3(5\mathbf{i} - \mathbf{j}) = 6\mathbf{i} + 4\mathbf{j} - 15\mathbf{i} + 3\mathbf{j} = -9\mathbf{i} + 7\mathbf{j}$.

Finding the magnitude and direction of a vector

Another advantage of expressing a vector in terms of **i** and **j** components is that it
is easy to find its magnitude and direction.

If $\mathbf{r} = a\mathbf{i} + b\mathbf{j}$ then the magnitude of **r** is given by

$$|\mathbf{r}| = \sqrt{a^2 + b^2}.$$

And the direction of r relative to i is given by

$$\theta = \tan^{-1}\left(\frac{b}{a}\right).$$

For example, if $\mathbf{r} = 3\mathbf{i} + 4\mathbf{j}$ then $|\mathbf{r}| = \sqrt{3^2 + 4^2} = 5$. and
$\theta = \tan^{-1}(\frac{4}{3}) = 53.1°$.

Progress check

1 Write down the position vector **r** of a point with coordinates (5, −2).

2

Given that $|\mathbf{r}| = 12$, express **r** in the
form $\mathbf{r} = a\mathbf{i} + b\mathbf{j}$.

3 Find the magnitude of the vector $\mathbf{r} = 5\mathbf{i} - 12\mathbf{j}$ and give its direction relative to **i**.

3 $|\mathbf{r}| = 13$, 67.4° clockwise.
2 $\mathbf{r} = 6\sqrt{3}\mathbf{i} + 6\mathbf{j}$
1 $\mathbf{r} = 5\mathbf{i} - 2\mathbf{j}$

62

3.2 Kinematics

After studying this section you should be able to:

- *apply the constant acceleration formulae for motion in a straight line*
- *draw and interpret graphs of displacement, velocity and acceleration against time*
- *use vectors to analyse motion in two and three dimensions*
- *apply the constant acceleration formulae to projectiles*

Motion in a straight line

You can use these formulae whenever the acceleration of an object is constant.

$v = u + at$	s is the displacement from a fixed position
$s = ut + \frac{1}{2}at^2$	u is the initial velocity
$s = \left(\dfrac{u+v}{2}\right)t$	a is the acceleration
	t is the time that the object has been in motion
$v^2 = u^2 + 2as$	v is the velocity at time t.

Example

An object starts from rest and moves in a straight line with constant acceleration 3 m s^{-2}. Find its velocity after 5 s.

From the given information: $u = 0$
$a = 3$
$t = 5$.

> Don't include units in your working.

> Make a clear statement and include the appropriate units.

Using $v = u + at$
gives $v = 0 + 3 \times 5 = 15$.

The velocity after 5 s is 15 m s^{-1}.

> **KEY POINT**
>
> Motion may take place in either direction along a straight line. One direction is taken to be positive for displacement, velocity and acceleration and the other direction is taken to be negative.

Example

> In some questions the acceleration due to gravity is taken to be 9.8 m s^{-2}.

A stone is thrown vertically upwards with a velocity of 20 m s^{-1}. It has a downward acceleration due to gravity of 10 m s^{-2}.

(a) Find the greatest height reached by the stone.

(b) Find its velocity after 3 s.

(c) Find the height of the stone above the point of projection after 3 s.

(a) From the given information: $u = 20$
$a = -10$.

At the greatest height $v = 0$.
Using $v = u + at$
$0 = 20 - 10t \Rightarrow t = 2$.

Using $s = ut + \frac{1}{2}at^2$
gives $s = 20 \times 2 - 5 \times 4 = 20$

The greatest height reached is 20 m.

(b) Using

$v = u + at$ with $t = 3$

gives

$v = 20 - 10 \times 3 = -10$.

After 3 s the stone is moving downwards at 10 m s^{-1}.

(c) Using

$s = ut + \frac{1}{2} at^2$

gives

$s = 20 \times 3 - 5 \times 3^2 = 15$.

After 3 s the stone is 15 m above the point of projection.

Graphical representation

You need to know the properties of the graphs of distance, displacement, speed, velocity and acceleration against time.

> The gradient of a distance–time graph represents speed.
> The gradient of a displacement–time graph represents velocity.
> The gradient of a velocity–time graph represents acceleration.
>
> The area under a speed–time graph represents the distance travelled.
> The area under a velocity–time graph represents the change of displacement.
> The area under an acceleration–time graph represents change in velocity.

KEY POINT

Example

The diagram represents the progress of a car as it travels between two sets of traffic lights.
Find:

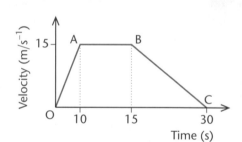

(a) The initial acceleration of the car.

(b) The deceleration of the car as it approaches the second set of lights.

(c) The distance between the traffic lights.

(a) Gradient of OA $= \frac{15}{10} = 1.5$
The initial acceleration of the car is 1.5 m s^{-2}.

(b) Gradient of BC $= -\frac{15}{15} = -1$.
The *acceleration* of the car is -1 m s^{-2} so the *deceleration* is 1 m s^{-2}.

(c) The area of the trapezium OABC is given by $\frac{15}{2}(5 + 30) = 262.5$,
so, the distance between the traffic lights is 262.5 m.

Using vectors

The work of this section may be extended to two and three dimensions using vectors.

Example

A particle has velocity $2\mathbf{i} - 3\mathbf{j}$ m s^{-1} when $t = 0$ and a constant acceleration of $\mathbf{i} + \mathbf{j}$ m s^{-2}. Find the speed of the particle when $t = 5$ and give its direction.

> The constant acceleration formulae work equally well in two or three dimensions.

From the given information: $\mathbf{u} = 2\mathbf{i} - 3\mathbf{j}$

$$\mathbf{a} = \mathbf{i} + \mathbf{j}$$

$$t = 5.$$

Using $\mathbf{v} = \mathbf{u} + \mathbf{a}t$ gives $\mathbf{v} = 2\mathbf{i} - 3\mathbf{j} + 5(\mathbf{i} + \mathbf{j}) = 7\mathbf{i} + 2\mathbf{j}$.

Speed is the magnitude of the velocity, giving $v = \sqrt{7^2 + 2^2} = 7.28\dots$.

So the speed of the particle when $t = 5$ is 7.28 m s^{-1} to 2 d.p.

From the diagram, $\tan \theta = \frac{2}{7} \implies \theta = 15.9°$.

When $t = 5$ the particle is moving at 15.9° anticlockwise to the direction of $\mathbf{i}$.

Progress check

1 A stone is thrown vertically upwards with a speed of 15 m s^{-1}. Take the downward acceleration due to gravity to be 10 m s^{-2} and find:
   (a) The greatest height reached.
   (b) The speed and direction of the stone after 2 s.

1 (a) 11.25 m (b) 5 m s^{-1} downwards

3.3 Statics and moments

After studying this section you should be able to:

- resolve a single force into perpendicular components
- resolve a system of forces in a given direction
- find the resultant of a system of forces
- solve problems involving friction
- apply the conditions for equilibrium of coplanar forces in simple cases
- find the moment of a force

LEARNING SUMMARY

Force

Force is a vector quantity that influences the motion of an object. It is measured in newtons (N). For example, the **weight** of an object is the force exerted on it by gravity. An object with a mass of m kg has weight mg N. **Tension, reaction** and **friction** are other examples of force and will be met in this section.

Resolving forces

Here are some examples of resolving forces into two components at right-angles to each other.

In each case the forces are represented in magnitude and direction by the sides of the triangle. The force to be resolved is *always* shown on the hypotenuse.

> Notice the directions of the forces shown by the arrows. The diagram represents vector addition of the components.

> **KEY POINT**
>
> These are **vector diagrams** showing the relationship between a force and its components.

In a **force diagram**, you can *replace* a force with a pair of components that are equivalent to it. This will often make it easier to produce the equations necessary to solve a problem.

> Take care not to show both the force *and* its components on a **force diagram** or you will represent the force TWICE.

An important example is that of an object on an **inclined plane**.

The diagram shows an object of weight W on a smooth plane inclined at angle θ to the horizontal. R is the force that the plane exerts on the object. It acts at right-angles to the plane and is called the **normal reaction**.

Force diagram

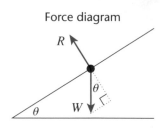

To analyse the behaviour of the object the weight is usually resolved into components parallel and perpendicular to the plane.

This vector diagram shows the component of weight acting down the plane is $W \sin \theta$ and the component perpendicular to the plane is $W \cos \theta$.

> It's worth remembering these results.

Vector diagram

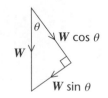

Example
An object of weight 10 N is held in place on a plane inclined at 30° to the horizontal by two forces F N and R N as shown.

Find the values of F and R.

The forces F and R correspond to the components of the weight, parallel and perpendicular to the plane, but act in the opposite directions.

> The forces must balance in each direction.

Resolving parallel to the plane: $F = 10 \sin 30° = 5$.

Resolving perpendicular to the plane: $R = 10 \cos 30° = 5\sqrt{3}$.

Resultant force

The effect of several forces acting at a point is the same as the effect of a single force called the **resultant force**.

You can find the resultant of a set of forces by using vector addition.

Example
Find the magnitude and direction of the resultant **R** N of the forces $(3\mathbf{i} - 2\mathbf{j})$ N, $(\mathbf{i} + 3\mathbf{j})$ N and $(-2\mathbf{i} + 4\mathbf{j})$ N.

The resultant force is given by: $\mathbf{R} = (3\mathbf{i} - 2\mathbf{j}) + (\mathbf{i} + 3\mathbf{j}) + (-2\mathbf{i} + 4\mathbf{j})$
$$= 2\mathbf{i} + 5\mathbf{j}.$$

The magnitude of the resultant is $|\mathbf{R}| = \sqrt{2^2 + 5^2} = \sqrt{29}$.

From the diagram: $\theta = \tan^{-1}(\tfrac{5}{2}) = 68.2°$

The resultant has magnitude $\sqrt{29}$ N and acts at 68.2° anticlockwise from the direction of **i**.

Equilibrium

A set of forces acting at a point is in **equilibrium** if the resultant force is zero.

It follows that the resolved parts of the forces must balance in any chosen direction.

Example
Find the force **F** such that the forces $(5\mathbf{i} - 3\mathbf{j})$, $(2\mathbf{i} + \mathbf{j})$ and **F** are in equilibrium.

For equilibrium $(5\mathbf{i} - 3\mathbf{j}) + (2\mathbf{i} + \mathbf{j}) + \mathbf{F} = 0$
$$\Rightarrow 7\mathbf{i} - 2\mathbf{j} + \mathbf{F} = 0$$
$$\Rightarrow \mathbf{F} = -7\mathbf{i} + 2\mathbf{j}.$$

Friction

Whenever two rough surfaces are in contact, the tendency of either surface to move relative to the other is opposed by the **force of friction**.

The diagram shows an object of weight W N resting on a rough horizontal surface. The object is pushed from one side by a force P N and friction responds with force F N in the opposite direction.

For small values of P, no movement takes place and $F = P$.

If P increases then F increases to maintain equilibrium until F reaches a maximum value. At this point the object is in **limiting equilibrium** and is on the point of slipping.

If P is now increased again then F will remain the same. The equilibrium will be broken and the object will move.

The maximum value of F depends on:

• The magnitude of the normal reaction R.
• The roughness of the two surfaces measured by the value μ. This value is known as the **coefficient of friction**.

In general: $F \leqslant \mu R$.
For limiting equilibrium: $F = \mu R$.

Example
An object of mass 8 kg rests on a rough horizontal surface. The coefficient of friction is 0.3 and a horizontal force P N acts on the object which is about to slide. Take $g = 10 \text{ m s}^{-2}$ and find the value of P.

> Remember that the weight of the object is given by mg.

Resolving vertically gives $R = 80$.

Friction is limiting so $F = \mu R = 0.3 \times 80 = 24$.

Resolving horizontally: $P = F \Rightarrow P = 24$.

Example
In the diagram, the object of mass m kg is on the point of slipping down the plane. The coefficient of friction between the object and the plane is μ. Show that $\mu = \tan \theta$.

Resolving perpendicular to the plane $\qquad R = mg \cos \theta$.
Resolving parallel to the plane $\qquad F = mg \sin \theta$.
Friction is limiting so $\qquad\qquad\qquad F = \mu R$.

This gives $\qquad\qquad\qquad mg \sin \theta = \mu mg \cos \theta \Rightarrow \mu = \dfrac{\sin \theta}{\cos \theta} = \tan \theta$.

Moments

The **moment of a force** about a point is a measure of the turning effect of the force about the point.
It is found by multiplying the force by its perpendicular distance from the point.
The moment of a force acts either clockwise ↻ or anticlockwise ↺ about the point.

The units of force (N) are multiplied by the units of distance (m) to give N m.

The moment of F about
P is Fd N m ↻.

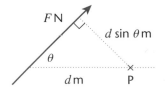

The moment of F about P
is $Fd \sin \theta$ N m ↻.

When there is more than one force, the **resultant moment** about a point is found by adding the separate moments. One direction is taken to be positive and the other negative.

Example   Find the resultant moment about O of the forces shown in the diagram.

Taking ↺ as positive:
The total moment is $8 \times 3 - 10 \times 2$ N m ↺

$$= 4 \text{ N m } ↺.$$

Equilibrium

For an object to be in equilibrium, the resultant force acting on it must be zero and any resultant moment must be zero.

In this context, a light rod is a rod is taken to have a mass that is so small that it can be ignored in any calculations.

Example
The diagram shows a light rod AB in equilibrium. Find the values of F and d.

Resolving vertically $F - 30 - 20 = 0$

$\Rightarrow$ $\qquad\qquad\qquad\qquad\quad F = 50$

Taking moments about C (written as M(C)) will give an equation involving d.

M(C) gives $\qquad 30d - 20 \times 1.5 = 0$

$\Rightarrow$ $\qquad\qquad\qquad\qquad 30d = 30$

$\Rightarrow$ $\qquad\qquad\qquad\qquad\quad d = 1.$

Example
If $F_1 = 10$N and $F_2 = 15$N, find F_3 and θ.

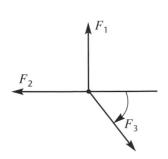

Resolving $\leftrightarrow F_2 = 15 = F_3 \cos \theta$

$\qquad\qquad \updownarrow F_1 = 10 = F_3 \sin \theta$

$\tan \theta = \dfrac{10}{15} \Rightarrow \theta = 33.7°$

$F_3 = \dfrac{15}{\cos \theta} = 18.0$N

69

Progress check

1 An object of mass 8 kg rests on a plane inclined at 40° to the horizontal. Find the components of its weight parallel and perpendicular to the plane. Take $g = 9.81$ m s^{-2}.

2 Find the magnitude and direction of the resultant **R** N of the forces $(2\mathbf{i} + 3\mathbf{j})$ N, $(5\mathbf{i} - 4\mathbf{j})$ N and $(-3\mathbf{i} - 7\mathbf{j})$ N.

3 An object of weight 50 N rests on a rough horizontal surface. A horizontal force of 20 N is applied to the object so that it is on the point of slipping. Find the value of the coefficient of friction.

1 parallel: 50.4 N, perp: 60.1 N
2 8.94 N at 63.4° ↻ to **i**
3 0.4

3.4 Dynamics

After studying this section you should be able to:

- understand and apply Newton's laws of motion in two or three dimensions
- analyse the motion of connected particles
- understand and apply the principle of conservation of momentum
- understand and use the coefficient of friction for moving particles
- use the relationship between impulse and momentum to solve problems

LEARNING SUMMARY

Newton's laws of motion

Newton's laws of motion provide us with a clear set of rules that can be used to analyse the effect of forces within a system.

The laws may be stated as:

1. Every particle continues in a state of rest or uniform motion, in a straight line, unless acted upon by an external force.

2. The resultant force acting on a particle is equal to its rate of change of momentum. *This law is most often applied in the form $F = ma$.*

3. Every force has an equal and opposite reaction.

The formula F = ma

In the formula $F = ma$, F N stands for the resultant force acting on a particle, m kg is the mass of the particle and a m s^{-2} is the acceleration produced.

It is important to use the correct units.

Example
A particle of mass 2 kg rests on a smooth horizontal plane.
Horizontal forces of 15 N and 4 N act on the particle in opposite directions.
Find the acceleration of the particle.

Using $F = ma$

$$15 - 4 = 2a$$

$$\Rightarrow \qquad a = 5.5$$

The acceleration of the particle is 5.5 m s^{-2}.

The formula applies equally well in two or three dimensions using vectors.

Example
Forces $(7\mathbf{i} - 2\mathbf{j} + \mathbf{k})$ N and $(3\mathbf{i} + \mathbf{j} - 5\mathbf{k})$ N act on a particle of mass 10 kg.
Find the acceleration produced.

The resultant force is the vector sum of the given forces, so using $\mathbf{F} = m\mathbf{a}$ gives

$$(7\mathbf{i} - 2\mathbf{j} + \mathbf{k}) + (3\mathbf{i} + \mathbf{j} - 5\mathbf{k}) = 10\mathbf{a}$$

$$\Rightarrow \qquad 10\mathbf{i} - \mathbf{j} - 4\mathbf{k} = 10\mathbf{a}$$

$$\Rightarrow \qquad \mathbf{a} = \mathbf{i} - 0.1\mathbf{j} - 0.4\mathbf{k}$$

The acceleration of the particle is $\mathbf{i} - 0.1\mathbf{j} - 0.4\mathbf{k}$ m s^{-2}.

The formula applies even when the force is variable.

Connected particles

In a typical problem, two particles are connected by a **light inextensible string**.
Since the string is light, there is no need to consider its mass.
Since it is inextensible, both particles will have the same speed and accelerate at the same rate *while the string is taut*.

By Newton's third law, the tension in the string acts equally on both particles but in opposite directions.

To solve problems involving connected particles you need to:

- Draw a diagram.
- Identify and label the forces acting on each particle.
- Write the **equation of motion** for each particle i.e. apply $F = ma$ for each one.
- Solve the simultaneous equations produced.

Example

Two particles P and Q are connected by a light inextensible string. The particles are at rest on a smooth horizontal surface and the string is taut. A force of 10 N is applied to particle Q in the direction PQ. P has mass 2 kg and Q has mass 3 kg. Find the tension in the string and the acceleration of the system.

> Note the use of the double headed arrow ➤➤ to represent acceleration.

P
T N
2 kg
T N
Q
3 kg
10 N
a m s^{-2}

| Using $F = ma$ | For particle P | $T = 2a$ | [1] |
| | For particle Q | $10 - T = 3a$ | [2] |

[1] + [2] gives $10 = 5a \Rightarrow a = 2$

Substituting for a in [1] gives $T = 4$.

The tension in the string is 4 N and the acceleration of the system is 2 m s^{-2}.

Momentum

The **momentum** of a particle is a vector quantity given by the product of its mass and its velocity i.e. momentum $= m\mathbf{v}$ where m kg is the mass of a particle and $\mathbf{v}$ m s^{-1} is its velocity.

The principle of conservation of momentum states that when two particles collide:

the total momentum before impact = the total momentum after impact.

> This kind of diagram gives a useful way to represent the information.

Before impact u_1 m s^{-1} u_2 m s^{-1}

A m_1 kg   B m_2 kg

After impact v_1 m s^{-1} v_2 m s^{-1}

> The positive direction for velocity is shown from left to right.

In the diagram, $u_1 > u_2$ so that the particles collide.

Conservation of momentum gives $m_1 u_1 + m_2 u_2 = m_1 v_1 + m_2 v_2$.

Example

A particle of mass 5 kg moving with speed 4 m s^{-1} hits a particle of mass 2 kg moving in the opposite direction with speed 3 m s^{-1}. After the impact the two particles move together with the same speed v m s^{-1}. Find the value of v.

Unknown velocities are shown in the positive direction.

Before impact

After impact

Conservation of momentum gives $5 \times 4 - 2 \times 3 = 5v + 2v$

$\Rightarrow \qquad 7v = 14$

$\Rightarrow \qquad v = 2.$

Impulse

The **impulse** of a constant force acting over a given time is given by the product of force and time. This may be written as $I = Ft$ where the impulse is I N s, the force is F N and the time is t s.

It follows that: Impulse = change in momentum

so $\qquad Ft = mv - mu$.

The total momentum of the system is conserved.

When two particles collide, each receives an impulse from the other of equal size but opposite sign. In this way, the total change in momentum is zero as expected.

Example

A particle of mass 4 kg, initially at rest, is acted upon by a force of 3 N for 10 s. What is the speed of the particle at the end of this time?

Using $\quad Ft = mv - mu$

gives $\quad 3 \times 10 = 4v - 0$

$\Rightarrow \qquad v = 7.5$

The speed of the particle is 7.5 m s^{-1}.

When two particles collide, the contact force between them may only last a very short time and is unlikely to be constant. However, the value of F in the equation $Ft = mv - mu$ may be used to represent the average value of this force.

Example

A ball of mass 0.5 kg strikes a hard floor with speed 2 m s^{-1} and rebounds with speed 1.5 m s^{-1}.

Given that the ball is in contact with the floor for 0.05 s find the average value of the contact force.

Taking upwards as the positive direction:

$u = -2$, $v = 1.5$, $m = 0.5$ and $t = 0.05$

Using $\quad Ft = mv - mu$

gives $\quad 0.05F = 0.5 \times 1.5 - 0.5 \times (-2)$

$\Rightarrow \qquad 0.05F = 1.75$

$\Rightarrow \qquad F = 35$

The average value of the contact force is 35 N.

Dynamic friction

When a particle moves over a rough surface, friction acts with its maximum value μR in the opposite direction to the motion.

In the diagram , $F = \mu R$

Progress check

1   A particle of mass 8 kg is acted upon by the forces $(11\mathbf{i} - 3\mathbf{j})$ N, $(14\mathbf{i} + \mathbf{j})$ N and $(-\mathbf{i} + 10\mathbf{j})$ N. Find the acceleration of the particle.

2   Two particles A and B are connected by a light inextensible string. The particles are at rest on a smooth horizontal surface and the string is taut. A force of 12 N is applied to particle B in the direction AB. A has mass 4 kg and B has mass 2 kg.
Find the tension in the string and the acceleration of the system.

3   A particle of mass 8 kg moving with speed 2 m s^{-1} hits a stationary particle of mass 2 kg.
After the impact both particles move in the same direction with speed v m s^{-1}.
(a) Find the value of v.
(b) Find the impulse given to the stationary particle.

4   A particle of mass 0.2 kg moves over a rough horizontal surface with initial speed 3 ms^{-1}. The coefficient of friction between the particle and the surface is 0.25. How far does the particle travel before it is brought to rest?

4 1.84 m (3 s.f.)
3 (a) $v = 1.6$ (b) Impulse = 3.2 N s.
2 Tension = 8 N, acceleration = 2 m s^{-2}.
1 $3\mathbf{i} + \mathbf{j}$ m s^{-2}.

Sample questions and model answers

1

A snowboarder pushes off with an initial speed of 3 m s^{-1}, in a straight line. After 5 seconds, he is travelling at 5 m s^{-1}.

(a) Find an expression for his speed t seconds after the start.

(b) Find an expression for the distance travelled in t seconds.

(c) The length of the slope is 400 m. What is his speed at the bottom?

(a) $v = u + at \Rightarrow 5 = 3 + 5a \Rightarrow a = 0.4 \text{ m s}^{-2}$

$v = 3 + 0.4t$

(b) $s = ut + \dfrac{1}{2}at^2 \Rightarrow s = 3t + \dfrac{0.4}{2}t^2$

$s = 3t + 0.2t^2$

(c) $v^2 = u^2 + 2as \Rightarrow v^2 = 3^2 + 2 \times 0.4 \times 400$

$= 9 + 320$

$= 329$

$v = \sqrt{329}$

$= 18.2 \text{ m s}^{-1}$

Sample questions and model answers *(continued)*

2

Two particles are connected by a light inextensible string passing over a light frictionless pulley. Particle A has mass 5 kg and lies on a smooth horizontal table. Particle B has mass 4.5 kg and hangs freely. Take $g = 10 \text{ m s}^{-2}$.

(a) Write down the equation of motion for each particle.

(b) Find the acceleration of the system.

(c) Find the tension in the string.

(a)

A clearly labelled diagram is an essential first step.

Both particles have the same acceleration.

The tension is the same at both ends of the string.

The diagram shows the forces acting on each particle in the direction of motion.

The equations of motion are:

Use $F = ma$ for each particle in turn.

for particle A $T = 5a$ [1]

for particle B $45 - T = 4.5a$ [2]

Solve the simultaneous equations and interpret the results.

(b) [1] + [2] gives $45 = 9.5a$

$$\Rightarrow a = 4.74 \text{ to 3 s.f.}$$

The acceleration of the system is 4.74 ms^{-2} to 3 s.f.

(c) Substitution for a in [1] gives $T = 23.7$ to 3 s.f.

It is a good idea to state the degree of accuracy given in your answer.

The tension in the string is 23.7 N to 3 s.f.

Practice examination questions

1 A particle moving in a straight line passes through a point O with velocity 10 m s^{-1} when $t = 0$. Given that the acceleration of the particle is -4 m s^{-2}.

Find:

(a) The velocity of the particle when $t = 5$.

(b) The distance that the particle travels between $t = 0$ and $t = 4$.

2

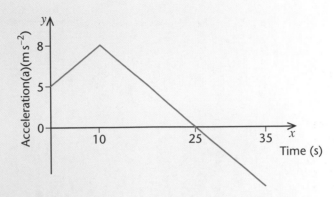

The diagram is an acceleration–time graph for the motion of a particle during a period of 35 s. The particle moves in a straight line and its initial velocity is 20 m s^{-1}.

(a) State the time at which the velocity reaches its maximum value.

(b) Calculate the maximum value of the velocity of the particle during this period.

3 A particle passes through a point P with position vector $(-5\mathbf{i} + 11\mathbf{j} + \mathbf{k})$ with velocity $(2\mathbf{i} + \mathbf{j})$ m s^{-1} when $t = 0$. The particle moves with constant acceleration $(\mathbf{i} + 3\mathbf{k})$ m s^{-2} and reaches the point Q when $t = 4$.

(a) Find $\overrightarrow{PQ}$.

(b) Find the position vector of Q.

(c) Find the average speed of the particle between P and Q.

4

The diagram shows an object of mass 12 kg resting on a rough horizontal surface.

A force of 50 N acts on the object at an angle of 40° to the horizontal.
Take $g = 10$ m s^{-2}.

(a) Find the normal reaction between the horizontal surface and the object.

(b) Given that the object is in limiting equilibrium, find the coefficient of friction between the object and the horizontal surface.

Practice examination questions *(continued)*

5 The diagram shows a mass of 6 kg acted upon by a horizontal force of *P* N.

The coefficient of friction between the object and the plane is 0.3.

Find the minimum value of *P* required to maintain the object in equilibrium.

6

Particles A and B of mass 2 kg and 3 kg respectively are connected by a light inextensible string passing over a smooth frictionless pulley. The system is released from rest with both particles 1 m above the ground. Find:

(a) The acceleration of the system while the string remains taut.

(b) The tension in the string.

(c) The time taken for particle B to reach the ground.

(d) The speed with which particle B hits the ground.

(e) The greatest height reached by particle A.

(f) What difference would it make to your initial equations if the pulley was not light and frictionless?

7

A particle P of mass $2m$ moving with speed $5u$ strikes particle Q of mass m, which is at rest.
After the collision both particles move in the same direction but the speed of particle Q is twice the speed of particle P.

(a) Find the speed of particle P after the collision.

(b) Find the magnitude of the impulse exerted on Q by P during the impact.

Chapter 4

Statistics 1

The following topics are covered in this chapter:

- Representing data
- Probability
- Discrete random variables

- The normal distribution
- Correlation and regression

4.1 Representing data

After studying this section you should be able to:

- calculate averages for discrete and continuous data
- find the variance, standard deviation, range and inter-quartile range of data
- use diagrams to compare distributions
- interpret measures of location and dispersion in comparing sets of data
- understand the concepts of outliers and skewness

LEARNING SUMMARY

Measures of central location

An average is a value that is taken to be representative of a data set. The three forms of average that you need are the mean, median and mode. These are often referred to as **measures of central location**.

The **mean** $\bar{x}$ of the values $x_1, x_2, x_3, \ldots, x_n$ is given by $\bar{x} = \dfrac{\sum x_i}{n}$. If each x_i occurs with frequency f_i then the mean of the **frequency distribution** is given by $\bar{x} = \dfrac{\sum f_i x_i}{\sum f_i}$.

Example
Find the mean of these results obtained by throwing a dice.

score	1	2	3	4	5	6
frequency	18	17	23	20	24	18

> One problem with the mean is that it may be unduly influenced by a small number of extreme values, known as **outliers**.

$$\bar{x} = \frac{18 \times 1 + 17 \times 2 + 23 \times 3 + 20 \times 4 + 24 \times 5 + 18 \times 6}{18 + 17 + 23 + 20 + 24 + 18} = \frac{429}{120} = 3.575$$

The **median** is the middle value of the data when it is arranged in order of size. If there is an even number of values then the median is the mean of the middle pair. In the example above, the 60th and 61st values are both 4 so the median is 4.

The **mode** is the value that occurs with the highest frequency. In the example above the mode is 5.

You can *estimate* the mean of **grouped data** by using the midpoint of each class interval to represent the class.

Example
Estimate the mean value of h from the figures given in the table.
An estimate of the mean is given by

$$\bar{x} = \frac{755}{72} = 10.486\ldots$$

$$10.5 = \text{to 1 d.p.}$$

> Once data have been grouped, the exact values are not available and so it is only possible to *estimate* the mean.

Interval	frequency (f_i)	midpoint (x_i)	$f_i \times x_i$
$0 < h \leqslant 5$	8	2.5	20
$5 < h \leqslant 10$	24	7.5	180
$10 < h \leqslant 15$	29	12.5	362.5
$15 < h \leqslant 20$	11	17.5	192.5
Totals	72		755

Measures of dispersion

An average alone gives no indication of how widely dispersed the data values are. The simplest measure of dispersion, or spread, is the **range**.

The range of a data set is the difference between largest value and the smallest value.

A slightly more refined approach is to measure the spread of the 'middle half' of the data. The data is put into order and the values Q1, Q2 and Q3 are found that divide the data into quarters.

> The diagram illustrates the situation for a data set of 11 values.

Q1   Q2   Q3

Q2 is the **median** of the data.

Q1, the **lower quartile**, is the median of the lower half of the data.

Q3, the **upper quartile**, is the median of the upper half of the data.

The **inter-quartile range** is then Q3 – Q1.

The quartiles may also be used to indicate whether the data values show **positive skew** (Q2 – Q1 < Q3 – Q2) or **negative skew** (Q2 – Q1 > Q3 – Q2).

The **variance** of a data set is the mean of the squares of the deviations from the mean.

> The second form is easier to work with from raw data.

$$\text{Variance} = \frac{\sum (x_i - \bar{x})^2}{n} = \frac{\sum x_i^2}{n} - \bar{x}^2.$$

The variance gives an indication of the spread of data values about the mean but the units of the data have been squared in the process.

> The value may be obtained directly from a scientific calculator but you may need to apply the formula.

$$\text{Standard deviation} = \sqrt{\text{variance}} = \sqrt{\frac{\sum x_i^2}{n} - \bar{x}^2}.$$

Standard deviation is measured in the same units as the data and is the value most commonly used to measure dispersion at this level.

For grouped data, take the mid-interval value for x.

In order to simplify calculations, data can be **coded**.

If a variable x is coded so that $y = a + bx$, then $\bar{y} = a + b\bar{x}$ and standard deviation $s_y = b \times s_x$

Statistical diagrams

The use of statistical diagrams may help in comparing distributions and revealing further information about the data.

A **box and whisker plot** for example shows the location and spread of a distribution at a glance.

A **back-to-back stem and leaf diagram** allows direct comparison of two sets of data to be made.

> Stem and leaf diagrams have the advantage over bar charts that details of the data are available.

The diagram gives a sense of location and spread for each data set.

The spread of marks is similar for the boys and the girls but the average for the girls is higher than for the boys.

boys		coursework marks		girls
	2	4	1 7	
	5 3	3	2 5 8 8	
5 4 4 2 1		2	4 4 7 9	
8 7 6 4		1	3 6 6	
	6 5	0	7	

2 | 4 means 42%   4 | 1 means 41%

Histograms are useful for illustrating grouped continuous data.
The area of a bar is proportional to the frequency in that interval.
The height of the rectangle is given by the **frequency density**, where

$$\text{frequency density} = \frac{\text{frequency}}{\text{width of interval}}.$$

The **modal class** is the interval with the greatest frequency density.
This is the highest bar on the histogram.

Joining with straight lines the mid points of the tops of the bars of a histogram gives a **frequency polygon**. Comparisons can be made by superimposing more than one frequency polygon on the same diagram.

In a **cumulative frequency diagram**, cumulative frequencies (running totals) are plotted against upper class boundaries of the intervals. The median, Q2, and quartiles, Q1 and Q3, can be estimated from the graph. For n items,

Q1 is the value 25% ($\frac{1}{4}n$) through the data,
Q2 is the value 50% ($\frac{1}{2}n$) through the data,
Q3 is the value 75% ($\frac{3}{4}n$) through the data.

Progress check

1   For the numbers 8, 14, 20, 1, 6, 6, 13, 12, 4,
    (i) find   (a) the mean   (b) the standard deviation   (c) the median
          (d) the interquartile range,
    (ii) draw a box and whisker plot.

2   Find the mean and standard deviation of this set of data:

x	0	1	2	3
f	4	6	7	3

3

Mass (g)	$20 \leqslant m < 40$	$40 \leqslant m < 50$	$50 \leqslant m < 55$	$55 \leqslant m < 60$	$60 \leqslant m < 90$
Frequency	10	12	10	7	9

For the above data
(a) estimate the mean and the standard deviation,
(b) draw a histogram and state the modal class,
(c) draw a cumulative frequency curve and use it to estimate
     (i) the median   (ii) the interquartile range.

4.2 Probability

After studying this section you should be able to:

- *use Venn diagrams to represent and interpret combined events*
- *use set notation to express and apply probability laws*
- *use tree diagrams to represent a problem*

LEARNING SUMMARY

Venn diagrams and probability laws

Venn diagrams provide a useful way to represent information about **sets** of objects.

The reason for mentioning them here is that the diagrams, and the notation used to express results, have a direct interpretation and application in probability theory.

In each diagram, the rectangle represents the set S of all objects under consideration.

A ∪ B is read as A union B.

The circles represent particular sets A and B of objects within the set S.

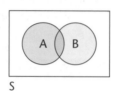

A ∪ B represents the set of all objects that belong to *either* A *or* B *or both*.

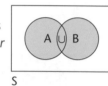

A ∩ B is read as A intersection B.

A ∩ B represents the set of those objects that belong to *both* A *and* B.

A′ represents the set of objects in S that do not belong to A.

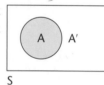

To make use of these ideas in probability theory:

- The objects are interpreted as **outcomes** for a particular situation.
- The set S is the **sample space** of all possible outcomes for the situation.
- The sets A and B are **events** defined by a particular choice of outcomes.
- P(A) for **example**, represents the probability that the event A occurs.
- P(A ∪ B) represents the probability that *either* A *or* B *or both* occurs.
- P(A ∩ B) represents the probability that *both* A *and* B occur.
- P(A′) represents the probability that A does *not* occur.

The **addition rule** states that

P(A ∪ B) = P(A) + P(B) − P(A ∩ B).

A and B are described as **mutually exclusive events** if they have no outcomes in common. These are events that cannot both occur at the same time.

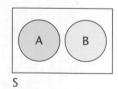

In this case P(A ∩ B) = 0 and the addition law becomes

$$P(A ∪ B) = P(A) + P(B).$$

P(A ∩ B) = 0

The events A and A′ are mutually exclusive for any event A.
It follows that P(A ∪ A′) = P(A) + P(A′).
Since one of the events A or A′ must occur, P(A ∪ A′) = 1 so P(A) + P(A′) = 1.

This is usually written as $P(A') = 1 - P(A)$.

The **multiplication rule** states that:

$$P(A \cap B) = P(A \mid B) \times P(B),$$

where $P(A \mid B)$ means the probability that A occurs *given that* B has *already* occurred.

$P(A \mid B)$ is described as a **conditional probability**, i.e. it represents the probability of A *conditional* upon B having occurred. Re-arranging the multiplication rule gives

Dependent

$$P(A \mid B) = \frac{P(A \cap B)}{P(B)}.$$

If A and B are **independent events** then the probability that either event occurs is not affected by whether the other event has already occurred.

In this case, $P(A \mid B) = P(A)$ and the multiplication rule becomes

$$P(A \cap B) = P(A) \times P(B).$$

Example
The events A and B are independent. $P(A) = 0.3$ and $P(B) = 0.6$
Find (a) $P(A \cup B)$ (b) $P(A' \cap B)$.

(a) $P(A \cup B) = P(A) + P(B) - P(A \cap B)$.

> In some problems you need to define the events first before you can apply the probability rules.

Since A and B are independent, $P(A \cap B) = P(A) \times P(B)$.

So $P(A \cup B) = 0.3 + 0.6 - 0.3 \times 0.6 = 0.72$

(b) A and B are independent $\Rightarrow$ A' and B are independent.

So $P(A' \cap B) = P(A') \times P(B) = (1 - 0.3) \times 0.6 = 0.42$

A **tree diagram** is a useful way to represent the probabilities of combined events. Each path through the diagram corresponds to a particular sequence of events and the multiplication rule is used to find its probability. When more than one path satisfies the conditions of a problem, these probabilities are added.

> This is true whether or not the events are independent.

Example
The probability that Chris gets grade A for Maths is 0.6 and the corresponding probability for English is 0.7. The events are independent.

(a) Calculate the probability that Chris gets just one A in the two subjects.
(b) Given that Chris gets just one grade A, find the probability that it is for Maths.

> Show the information on a tree diagram.

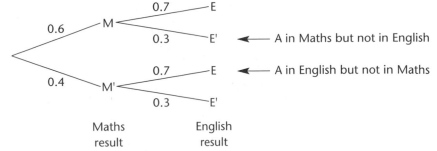

(a) $P(\text{just one A}) = 0.6 \times 0.3 + 0.4 \times 0.7$

> Multiply the probabilities along the branches.

$$= 0.18 + 0.28$$
$$= 0.46$$

Use conditional probability.

(b) P(A in Maths | just one A) = $\dfrac{\text{P(A in Maths and just one A)}}{\text{P(just one A)}}$

$= \dfrac{0.18}{0.46}$

$= 0.39$ (2 d.p.)

Progress check

1 The probability that the event B occurs is 0.7. The probability that events A and B both occur is 0.4. What is the probability that A occurs given that B has already occurred?

2 It is given that P(A) = 0.4, P(B) = 0.7 and P(A ∩ B) = 0.2
Find (a) P(A ∪ B)   (b) P(B | A)   (c) P(A | B).

3 A bag contains 3 red counters and 7 green counters. Two counters are taken at random from the bag, without replacement.
(a) By drawing a tree diagram, or otherwise, find the probability that the counters are different colours.
(b) Given that the counters are different colours, find the probability that the first counter picked is red.

3 (a) $\frac{7}{15}$ (b) $\frac{1}{2}$

2 (a) 0.9 (b) 0.5 (c) $\frac{2}{7}$

1 4/7

4.3 Discrete random variables

After studying this section you should be able to:

- understand the concept of a discrete random variable, its mean and variance
- calculate the expectation and variance of functions of a random variable
- understand the discrete uniform distribution

Discrete random variables

The score obtained when a dice is thrown may be thought of as a **random variable**. Since only specific values may be obtained, this is an example of a **discrete random variable**. In the usual notation:

- A capital letter such as X is used as a label for the random variable.
- A lower case letter such as x is used to represent a particular value of X.
- The probability that X takes the value x is written as $P(X = x)$ or just $p(x)$ and this is known as the **probability function**.

> This is a useful result.

- If X is a discrete random variable with probability function $p(x)$ then $\sum p(x) = 1$.

The **probability distribution** of X is given by the set of all possible values of x together with the values of $p(x)$. This is usually shown in a table.

For example, the probability distribution of the scores shown on a dice may be given as:

x:	1	2	3	4	5	6
$p(x)$:	$\frac{1}{6}$	$\frac{1}{6}$	$\frac{1}{6}$	$\frac{1}{6}$	$\frac{1}{6}$	$\frac{1}{6}$

> This is an example of a discrete *uniform* distribution, as all the probabilities are the same.

The **cumulative distribution function** is given by $F(x_0) = P(X \leqslant x_0) = \sum_{x \leqslant x_0} p(x)$.

This represents the probability that the random variable X takes a value less than or equal to x_o and is given by the sum of the probabilities up to and including that point.

Mean and variance

The symbol μ is used to stand for the **mean value of X**. This is also known as the **expected value, or expectation, of X** and is written $E(X)$.

> $E(X) = \sum x P(X = x)$

For any discrete random variable X the mean is $\mu = E(X) = \sum x p(x)$.

Example Find the expected value of the score shown on a dice.

x:	1	2	3	4	5	6
$p(x)$:	$\frac{1}{6}$	$\frac{1}{6}$	$\frac{1}{6}$	$\frac{1}{6}$	$\frac{1}{6}$	$\frac{1}{6}$
$x p(x)$:	$\frac{1}{6}$	$\frac{2}{6}$	$\frac{3}{6}$	$\frac{4}{6}$	$\frac{5}{6}$	$\frac{6}{6}$

$$\mu = \frac{1}{6} + \frac{2}{6} + \frac{3}{6} + \frac{4}{6} + \frac{5}{6} + \frac{6}{6} = \frac{21}{6}$$

The expected value is 3.5

$E(g(X)) = \Sigma g(x) P(X = x)$

$E(X^2) = \Sigma x^2 P(X = x)$

The **expected value of a function of a random variable** is found in a similar way. In general, if $g(X)$ is some function of the random variable X then

$$E(g(X)) = \Sigma g(x)p(x).$$

For example, $E(X^2) = \Sigma x^2 p(x)$. In the case of the dice scores above this gives:

$$E(X^2) = \tfrac{1}{6} + \tfrac{4}{6} + \tfrac{9}{6} + \tfrac{16}{6} + \tfrac{25}{6} + \tfrac{36}{6} = \tfrac{91}{6}.$$

Using σ to stand for the **standard deviation of X** and Var(X) to stand for the **variance of X** we have:

$$\sigma^2 = \text{Var}(X) = E(X^2) - \mu^2.$$

Referring to the example of the dice scores again:
$\text{Var}(X) = \tfrac{91}{6} - \left(\tfrac{21}{6}\right)^2 = \tfrac{35}{12}$ and $\sigma = \sqrt{\tfrac{35}{12}}$.

The **expectation of a linear function of X** can be expressed in terms of $E(X)$.

In general: $E(aX + b) = aE(X) + b.$

Once the value of $E(X)$ is known, applying the result above is much simpler than working out $\Sigma (ax + b)p(x)$. However, the result only holds for linear functions, so for example $E(X^2)$ is not the same as $(E(X))^2$.

The **variance of a linear function of X** may be expressed in terms of Var(X).

In general: $\text{Var}(aX + b) = a^2 \text{Var}(X).$

Progress check

1 The random variable X has the probability distribution shown. x: 0 1 2
   Find the value of k. $p(x)$: 0.3 0.2 k

2 Find (a) $E(X)$ (b) Var(X) for the probability distribution in question 1.

3 The random variable Y is given by $Y = 3X + 2$, where X is as in question 1.
   Find (a) $E(Y)$ (b) Var(Y).

4 X has a discrete uniform distribution where $P(X = x) = k$ for $x = 30, 40, 50, 60, 70$.
   Find the value of k and the mean and standard deviation of X.

4 0.2, 50, 14.1
3 (a) 5.6 (b) 6.84
2 (a) 1.2 (b) 0.76
1 $k = 0.5$

4.4 *The normal distribution*

After studying this section you should be able to:

- *use the normal distribution model to calculate probabilities*

The normal distribution

The **normal variable** is a continuous random variable. Its probability function is represented by a bell-shaped curve, symmetric about the mean value μ.

The parameters of the normal distribution are μ and σ^2. If the continuous random variable X follows a normal distribution with these **parameters** then this is written as $X \sim N(\mu, \sigma^2)$. The standard deviation is σ.

Tabulated values of probabilities for the normal distribution use Z as the variable.

The **standard normal variable** Z is used to calculate probabilities based on the Normal distribution where $Z = \dfrac{X - \mu}{\sigma}$ and $Z \sim N(0, 1)$.

Example
Find $P(15 < X < 17)$ where $X \sim N(15, 25)$

$$15 < X < 17 \Rightarrow \frac{15 - 15}{5} < Z < \frac{17 - 15}{5} \Rightarrow 0 < Z < 0.4$$

Using the normal distribution tables.

$$P(15 < X < 17) = P(0 < Z < 0.4)$$
$$= \Phi(0.4) - \Phi(0) = 0.6554 - 0.5000$$
$$= 0.1554$$

You may need to use given information to set up and solve simultaneous equations to find the values of μ and σ.

Example
It is given that $P(X > 70) = 0.0436$ and $P(X < 45) = 0.3669$, where $X \sim N(\mu, \sigma^2)$. Find the values of μ and σ.

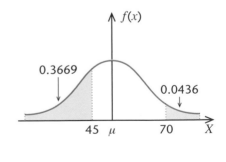

$$P(X > 70) = P\left(Z > \frac{70 - \mu}{\sigma}\right) = 0.0436$$

$$\Phi\left(\frac{70 - \mu}{\sigma}\right) = 1 - 0.0436 = 0.9564$$

From the tables

$$\left(\frac{70 - \mu}{\sigma}\right) = 1.71 \Rightarrow 70 - \mu = 1.71\sigma \qquad [1]$$

Similarly

$$P(X < 45) = P\left(Z < \frac{45 - \mu}{\sigma}\right) = 0.3669$$

$$\Phi\left(\frac{45 - \mu}{\sigma}\right) = 0.3669$$

In general,
$\Phi(-z) = 1 - \Phi(z)$.

$$\Phi\left(-\left(\frac{45 - \mu}{\sigma}\right)\right) = 1 - 0.3669 = 0.6331$$

$$-\left(\frac{45 - \mu}{\sigma}\right) = 0.34$$

$$45 - \mu = -0.34\sigma \qquad [2]$$

[1] − [2] gives $25 = 2.05\sigma \Rightarrow \sigma = 12.2$

Substituting in (1): $\mu = 49.1$

Progress check

1 Find P(17 < X < 24) where $X \sim$ N(14, 16).

2 It is given that $X \sim$ N(μ, σ^2) and P(X < 30) = 0.7.
  Given also that P(X > 35) = 0.1, find μ and σ.

1 0.2204
2 $\mu = 26.5$
   $\sigma = 6.60$

4.5 Correlation and regression

After studying this section you should be able to:

- *calculate and interpret the product-moment correlation coefficient*
- *recognise explanatory and response variables*
- *find the equations of least squares regression lines*
- *use a suitable regression line to estimate a value*

All of the work in this section relates to the treatment of **bivariate data**. This is data in which each data point is defined by two variables.

Correlation

A **scatter diagram** may be used to represent bivariate data. The extent to which the points approximate to a straight line gives an indication of the strength of a linear relationship between the variables, known as the **linear correlation**.

One way to arrive at a numerical measure of the correlation is to use the **product–moment correlation coefficient**, r.

You should only use this method when both variables are normally distributed.

For n pairs of (x, y) values:

$$S_{xx} = \Sigma(x - \bar{x})^2 = \Sigma x^2 - \frac{(\Sigma x)^2}{n} \qquad S_{yy} = \Sigma(y - \bar{y})^2 = \Sigma y^2 - \frac{(\Sigma y)^2}{n}$$

$$S_{xy} = \Sigma(x - \bar{x})(y - \bar{y}) = \Sigma xy - \frac{(\Sigma x)(\Sigma y)}{n}$$

and the product–moment correlation coefficient is given by:

You may be able to obtain the value of r directly from your calculator.

$$r = \frac{S_{xy}}{\sqrt{S_{xx}S_{yy}}}$$

This gives values of r between –1 (representing a perfect negative correlation) and +1 (representing a perfect positive correlation).

Regression

Whereas correlation is determined by the *strength* of a linear relationship between the two variables, **regression** is about the *form* of the relationship given by the equation of a **regression line**. (Make sure you known the difference between correlation and regression.)

The purpose in establishing the equation of a regression line is to make predictions about the values of one variable (known as the **response variable**) for some given values of the other variable (known as the **explanatory variable**).

Predictions should only be made for values within the range of readings of the explanatory variable. **Extrapolation** for values outside this range is unreliable. Another factor affecting the accuracy of any predictions is the influence of **outliers** on the equation of the regression line.

Figure 1 illustrates *y*-**residuals** given by

$$d = \text{(observed value of } y) - \text{(predicted value of } y).$$

Figure 2 illustrates *x*-**residuals** given by

$$d = \text{(observed value of } x) - \text{(predicted value of } x).$$

A **least squares regression line** is a line for which the sum of the squares of either the x-residuals or y-residuals is minimised.

Figure 1

Figure 2

Both lines always pass through the point $(\bar{x}, \bar{y})$.

$(\bar{x}, \bar{y})$ ╳

d

$\Sigma d^2 = \min$

$(\bar{x}, \bar{y})$ ╳

d

$\Sigma d^2 = \min$

This gives the regression line of y on x as

$$y = a + bx,$$

where $b = \dfrac{S_{xy}}{S_{xx}}$ and $a = \bar{y} - b\bar{x}$.

The values of a and b may be obtained directly from some calculators.

Use this equation to estimate values of y for given values of x when x is the explanatory variable and y is the response variable.

This gives the regression line of x on y as

$$x = c + dy,$$

where $d = \dfrac{S_{xy}}{S_{yy}}$ and $c = \bar{x} - d\bar{y}$.

Use this equation to estimate values of x for given values of y when y is the explanatory variable and x is the response variable.

Unless there is perfect correlation between the variables, the two regression lines will be different and you cannot rearrange one equation to obtain the other.

Progress check

The summary data for 10 pairs of (x, y) values is as follows:

$$\Sigma x = 146, \quad \Sigma x^2 = 2208, \quad \Sigma y = 147, \quad \Sigma y^2 = 2247, \quad \Sigma xy = 2211.$$

1 Find the value of the product–moment correlation coefficient between x and y.
2 Find the equation of the least squares regression line of y on x.

2 $y = 2.317 + 0.848x$
1 0.799

Sample questions and model answers

1

Paul swims 20 lengths of a swimming pool in training for a competition. His times, x seconds, for completing lengths of the pool are summarised by:

$$\sum x = 372 \qquad \sum x^2 = 6952.$$

(a) Find the mean and standard deviation of Paul's times for completing one length of the pool.

(b) A month ago, Paul's mean time per length was 19.8 s with a standard deviation of 1.72 s. Comment on the change in his performance indicated by these results.

Scientific calculators can find the standard deviation from the raw data, but you need to know the formulae because the data may be given in summarised form.

(a)
$$\bar{x} = \frac{\sum x}{n} = \frac{372}{20}$$

$$= 18.6$$

Paul's mean time per length is 18.6 s

$$\sqrt{\frac{\sum x^2}{n} - \bar{x}^2} = \sqrt{\frac{6952}{20} - 18.6^2}$$

$$= 1.28$$

Quote the formula and substitute the information.

The standard deviation of Paul's times is 1.28 s

(b) The mean time has been reduced, suggesting an improvement in performance. The standard deviation has also been reduced which suggests a greater consistency of performance.

You may be expected to interpret results within the context of the question and make comparisons.

2

A machine produces nails with lengths that are normally distributed with mean 5.2 cm and standard deviation 0.06 cm.

Find the probability that a nail selected at random has a length between 5.1 cm and 5.25 cm.

Define the random variable X.

Let X be the length, in cm, of a nail, where X ~ N (5.2, 0.06^2)

Express the probability in terms of X and then standardise.

$$P(5.1 < X < 5.25) = P\left(\frac{5.1-5.2}{0.06} < Z < \frac{5.25-5.2}{0.06}\right)$$

$$= P(-1.67 < Z < 0.83)$$

$$= \Phi(0.83) - \Phi(-1.67)$$

It follows from the symmetry of the distribution that

$$= \Phi(0.83) - (1 - \Phi(1.67))$$

$$\Phi(-z) = 1 - \Phi(z).$$

$$= 0.7967 - (1 - 0.9525)$$

$$= 0.7492$$

The probability that a nail selected at random has a length between 5.1 cm and 5.25 cm is 0.749 (3 s.f.)

1 A discrete random variable X has the probability function $p(x)$ given by

$$p(x) = \frac{x^2}{k} \quad \text{for} \quad x = 1, 2, 3.$$

$$p(x) = 0 \quad \text{otherwise.}$$

(a) Find the value of k.

(b) Calculate the mean and variance of the probability distribution.

(c) Find the mean and variance of the random variable $Y = 3X - 5$.

2 The events A and B are such that

$$P(A) = 0.3 \qquad P(B) = 0.6 \qquad P(A \cup B) = 0.72$$

(a) Show that the events A and B are not mutually exclusive.

(b) Show that the events A and B are independent.

(c) Write down $P(A \cup B)'$.

(d) Find $P(A' \cup B')$.

3 (a) Three cards are drawn at random, without replacement, from a pack of playing cards.
   Calculate the probability that:

   (i) All three cards are red.
   (ii) All three cards are the same colour.
   (iii) At least one of the cards is red.

   (b) The cards are returned to the pack and the pack is shuffled. One more card is drawn.
   Find the probability that this card is an ace, given that it is not a diamond.

4 The lifetime of batteries used in a torch can be modelled by a Normal distribution with a mean of 8 h and a standard deviation of 25 mins.

(a) Calculate the probability that a battery selected at random will last for more than $8\frac{1}{2}$ h.

(b) Find the percentage of batteries that will last for less than 7 h.

5 Eleven students in the sixth form of a school study both mathematics and physics at A Level. It is thought that their physics scores depend on their maths scores. The scores in an end of term test are shown for 10 of these students.

Maths (x)	48	39	52	55	71	62	76	54	61	58
Physics (y)	41	35	50	47	63	48	70	45	46	39

Using x to represent the score in maths and y to represent the corresponding score in physics:

$$\Sigma x = 576, \ \Sigma y = 484, \ \Sigma x^2 = 34216, \ \Sigma y^2 = 24450, \ \Sigma xy = 28785$$

(a) Calculate the product–moment correlation coefficient, between x and y.

(b) Find the equation of the regression line of y on x for the data.

(c) The eleventh student obtained 65 on the maths test. Estimate this student's score in the physics test, commenting on the reliability of your estimate.

Chapter 5

Decision Mathematics 1

The following topics are covered in this chapter:

- *Algorithms*
- *Graphs and networks*
- *Critical path analysis*

- *Linear programming*
- *Matchings*

5.1 Algorithms

After studying this section you should be able to:

- *implement an algorithm given by text or flow chart*
- *implement the following algorithms from memory:*
  - *bubble sort*
  - *quick sort*
  - *binary search*
  - *bin packing*

LEARNING SUMMARY

Introducing algorithms

An **algorithm** is a finite sequence of precise instructions used to solve a problem.

One way to define an algorithm is to simply list all of the necessary steps. In more complex situations, the process may be easier to follow if the algorithm is presented as a **flow diagram**.

Algorithms play a fundamental role throughout Decision Mathematics.

You need to be able to implement an algorithm presented in either form.

In the exam you may be asked to implement an unfamiliar algorithm and interpret the results.

Example
Implement the algorithm given by the flow chart and state what the written values represent.

The table shows how the values of I and J change as the algorithm is implemented.

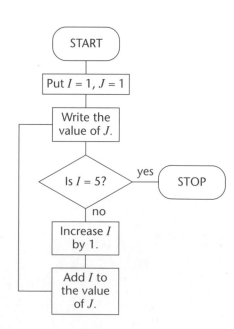

I	1	2	3	4	5
J	1	3	6	10	15

The written values are 1, 3, 6, 10, 15 and these are the first five triangle numbers.

The algorithm may be expressed in words as:

Write down a sequence of 5 numbers starting with 1 and then increasing, first by 2 and then by 1 more each time from term to term.

This version is more concise but not so easy to understand.

In the exam you may be expected to implement any of these algorithms without being reminded of the necessary steps.

You need to know and be able to implement the following algorithms.

The bubble sort

As its name suggests, the **bubble sort** is an algorithm for sorting a list in a particular order.

Step 1 Compare the first two elements of the list and switch them if they are in the wrong order.

Step 2 Compare the second and third elements of the list and, again, switch them if they are in the wrong order.

Step 3 Continue in the same way until you reach the end of the list. This completes the first **pass** through the list.

Step 4 Make repeated passes through the list until a pass produces no change.

Example Sort the list 5, 7, 3, 11, 6, 8 into ascending order.

The first pass gives

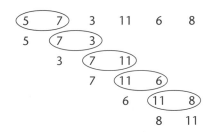

The second pass produces   3, 5, 6, 7, 8, 11.

The third pass produces   3, 5, 6, 7, 8, 11.

Don't just look at the list and say that the numbers are now in order. Use the algorithm to decide when to stop.

The third pass made no change so the ordering is complete.

The quick sort

This algorithm relies on finding the middle element of a list. For n elements the middle element has position $\frac{1}{2}(n + 1)$ when n is odd and $\frac{1}{2}(n + 2)$ when n is even.

Step 1 Take the middle element as the **pivot**.

Step 2 Consider each of the other elements in turn and place any with order less than or equal to the pivot on its one side and any greater than the pivot on its other side. Do this without re-ordering the numbers on either side of the pivot.

This creates two sub-lists with the pivot in-between.

Step 3 Repeat steps 1 and 2 for each sub-list until each one contains a single element.

Example Sort the list 4, 2, 6, 10, 5, 7 into ascending order.

Values shown highlighted have been used as pivots.

It is easy to see that the list is now in order but the sub-list 4, 5 has more than one element, so you must continue with the algorithm.

The ordered list is 2, 4, 5, 6, 7, 10.

Binary search

The binary search algorithm provides a systematic way of searching through an ordered list to find an element that matches your criterion. **For example**, you may need to look through a set of index cards to find the contact details of a customer.

> This is not a very efficient method if the required element is at the end of the list.

Step 1 Find the element in the middle of the list. If this element matches your criterion then stop. If it does not, then use the middle term to decide in which half of the list the required element must lie.

Step 2 Repeat Step 1 for the selected half.

Bin packing

The bin packing algorithm is used to solve problems that can be represented by the need to pack some boxes of equal cross-section but different heights into bins, with the same cross-section as the boxes, using as few bins as possible.

There is no known algorithm that will always provide the *best* solution. You need to be familiar with the three algorithms below that attempt to provide a *good* solution. Such algorithms are known as **heuristic** algorithms.

First-fit algorithm

Take each box in turn from the order given and pack it into the first available bin.

Full-bin algorithm

> The full-bin algorithm is only practical when the number of bins and boxes is small.

Step 1 Considering the boxes in the given order, use the first available combination that will fill a bin.

Step 2 Repeat Step 1 until no more bins can be filled.

Step 3 Implement the first-fit algorithm for the remaining boxes.

First-fit decreasing algorithm

Step 1 Arrange the boxes in decreasing order of size.

Step 2 Implement the first-fit algorithm starting with the largest box.

Progress check

1 Write the letters P, Q, B, C, A, D in alphabetical order using:

(a) the bubble sort algorithm

(b) the quick sort algorithm.

2 A project involves activities A–H with durations in hours as given in the table.

A	B	C	D	E	F	G	H
3	1	5	4	2	3	4	2

The project is to be completed in 8 hours.

(a) Use the first-fit algorithm to try to find the minimum number of workers needed.

(b) Find a better solution.

> You can represent each worker as a 'bin' of 'height' 8 hours.

1 A, B, C, D, P, Q.

2 First-fit: 4 workers; A better solution is (A, B, D) (C, F) (E, G, H).

5.2 Graphs and networks

After studying this section you should be able to:

- *understand the use of terminology associated with graphs and networks*
- *use Prim's and Kruskal's algorithms for constructing a minimum spanning tree*
- *use Dijkstra's algorithm to find a shortest path between two vertices*
- *use the route inspection algorithm to find a route of minimum weight that traverses every arc of a network*

LEARNING SUMMARY

Defining terms

It is worth spending time getting to know all of the terms. Their definitions are quite detailed so you will need to keep reminding yourself of what each term means.

- A **graph** is a set of points, called **vertices** or **nodes**, connected by lines called **edges** or **arcs**.
- A **simple graph** is one that has no loops and in which no pair of vertices are connected by more than one edge.

A simple graph

A non-simple graph

Two edges between A and B

A non-simple graph

A loop around C

- The number of edges incident on a vertex is called its **order**, **degree** or **valency**. A vertex may be **odd** or **even** depending on whether its order is odd or even.
- A **subgraph** of some graph G is a graph consisting entirely of vertices and edges that belong to G.
- A **directed** edge is an edge that has an associated direction shown by an arrow. A **digraph** is a graph in which the edges are directed.
- A **path** is a finite sequence of edges such that the end vertex of one edge is the start vertex of the next edge. No edge is included more than once.
- A **connected graph** is one in which every pair of vertices is connected by a path.
- A **complete graph** is one in which every pair of vertices is connected by an edge. A complete graph with n vertices is denoted by K_n.
- A **planar graph** is one that can be drawn in a plane such that no two edges meet except at a vertex.
  K_4 is planar but K_5 is not planar.

K_4 K_5

- A **cycle** is a path that starts and finishes at the same vertex.
- An **Eulerian** cycle is one that traverses all of the edges of the graph.
- A **tree** is a graph with no cycles.
- A **spanning tree** is a tree whose vertices are all of the vertices of the graph.
- A **network** is a graph that has a number (**weight**) associated with every edge.
- A **minimum spanning tree** (MST) of a network is a spanning tree of minimum possible weight. It is sometimes called a **minimum connector**.

Prim's algorithm

Prim's algorithm may be used to find a minimum spanning tree of a network.

Step 1 Choose a starting vertex.

Step 2 Connect it to the vertex that will make an edge of minimum weight.

Step 3 Connect one of the remaining vertices to the tree formed so far, in such a way that the minimum extra weight is added to the tree.

Step 4 Repeat step 3 until all of the vertices are connected.

> The number in row A and column B, for example, represents the weight of the edge AB.

The network given above may be represented by the matrix shown.

A version of Prim's algorithm may be used to find the minimum spanning tree directly from a matrix.

Step 1 Choose a starting vertex. Delete the corresponding row and write 1 above the corresponding column as a label.

Step 2 Circle the smallest undeleted value in the labelled column and delete the row in which it lies.

Step 3 Label the column, corresponding to the vertex of the deleted row, with the next label number.

> If there is more than one smallest value then you can choose which one to circle.

Step 4 Circle the smallest undeleted value of all the values in the labelled columns and delete the row in which it lies.

Step 5 Repeat steps 3 and 4 until all of the rows are deleted. The circled values then define the edges of the minimum spanning tree.

Starting from A gives the MST as

AD could have been used instead of CD

```
      A   B   C   D
A  –   5   –   –
B  5   –   7   –
C  –   7   –   14
D  –   –   14  –
```

```
         1   2   3
      A   B   C   D
A  –   5   –   –
B  (5) –   7   –
C  –   (7) –   14
D  –   –   (14) –
```

The circled values correspond to the edges AB, BC, and CD.

Kruskal's algorithm

Kruskal's algorithm is an alternative way to find a minimum spanning tree. In this case, the subgraph produced may not be connected until the final stage.

Step 1 Choose an edge with minimum weight as the first subgraph.

Step 2 Find the next edge of minimum weight that will not complete a cycle when taken with the existing subgraph. Include this edge as part of a new subgraph.

Step 3 Repeat step 2 until the subgraph makes a spanning tree.

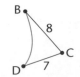

BD can not be included

Dijkstra's algorithm

Dijkstra's algorithm is used to find the shortest distance between a chosen *start* vertex and any other vertex in a network.

The algorithm is designed for use by computers but, when implementing it without the aid of a computer, a system of labelling is required.

The word *distance* is used here in place of the more general *weight* because in practical applications of Dijkstra's algorithm the weight so often represents a distance.

Reading the letters inside the box as OWL might help you to remember which part is for which information.

Order of labelling

Label (the minimum distance from the start vertex).

Record your 'Working values' here

The steps in the algorithm below refer to this system of labelling. Every vertex is labelled in the same way.

Step 1   Give the start vertex a label of 0 and write its order of labelling as 1.

Step 2   For each vertex directly connected to the start vertex, enter the distance from the start vertex as a working value. Enter the smallest working value as a label for that vertex. Record the order in which it was labelled as 2.

Step 3   For each vertex directly connected to the one that was last given a label, add the edge distance onto the label value to obtain a total distance. This distance becomes the working value for the vertex unless a lower one has already been found.

Step 4   Find the vertex with the smallest working value not yet labelled and label it. Record the order in which it was labelled.

Step 5   Repeat steps 3 and 4 until the target vertex is labelled. The value of the label is the minimum distance from the start vertex.

Step 6   To find the path that gives the shortest distance, start at the target vertex and work back towards the start vertex in such a way that an edge is only included if its distance equals the change in the label values.

Example
Use Dijkstra's algorithm to find the shortest distance from A to F in this network and state the route used.

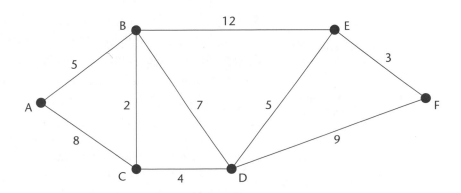

Using Dijkstra's algorithm produces the diagram below.

The shortest distance from A to F is 19 (given by the label at F).

Tracing the route backwards using Step 6 gives F, E, D, C, B, A, so the required route is A, B, C, D, E, F.

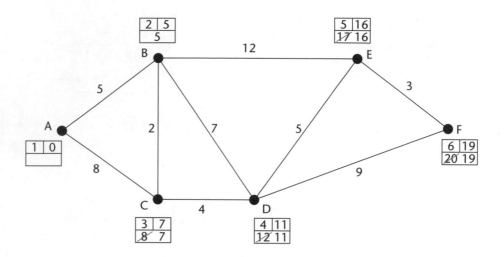

The route inspection problem

Again, it is common to use length rather than weight in this context.

The **route inspection problem** is to find a route of minimum total length that traverses every edge of the network, at least once, and returns to the start vertex.

This is also known as the Chinese postman problem.

The algorithm for solving this problem is based on the idea of a **traversable** graph. A graph is traversable if it can be drawn in one continuous movement without going over the same edge more than once. If all of the vertices of the graph are even, then any one of them may be used as the starting point and the same point will be returned to at the end of the movement. Such a graph is said to be **Eulerian**. The only other possibility for a traversable graph is that it has exactly two odd vertices. In this situation, one of the vertices is the start point and the other is the finish point. This type of graph is said to be **Semi-Eulerian**.

Leonhard Euler was the most prolific mathematician of all time. One of his many contributions to the subject was to introduce graph theory as a means of solving problems.

The **route inspection algorithm** is:

Step 1   List all of the odd vertices.

Step 2   Form the list into a set of pairs of odd vertices. Find all such sets.

Step 3   Choose a set. For each pair find a path of minimum length that joins them. Find the total length of these paths for the chosen set.

Step 4   Repeat steps 3 until all sets have been considered.

Repeating the edges between the pairs of odd vertices in this way effectively makes all the vertices even and creates a traversable graph.

Step 5   Choose the set that gives the minimum total. Each pair in the set defines an edge that must be repeated in order to solve the problem.

Progress check

1 (a) Use Prim's algorithm to find the total weight of a minimum spanning tree for this network.

(b) Verify your answer to part (a) using Kruskal's algorithm.

2 Use Dijkstra's algorithm to find the shortest distance from A to G in the network given in question 1. State the route used.

5.3 Critical path analysis

After studying this section you should be able to:

- *construct an activity network from a given precedence table including the use of dummies where necessary*
- *use forward and backward scans to determine earliest and latest event times*
- *find the critical path in an activity network*
- *calculate the total float of an activity*
- *construct a chart for the purpose of scheduling*

LEARNING SUMMARY

The process of representing a complex project by a network and using it to identify the most efficient way to manage its completion is called **critical path analysis**.

Activity networks

A complex project may be divided into a number of smaller parts called **activities**. The completion of one or more activities is called an **event**.

Activities often rely on the completion of others before they can be started.

The relationship between these activities can be represented in a **precedence table**, sometimes called a **dependency table**.

In the precedence table shown on the right, the figures in brackets represent the **duration** of each activity, i.e. the time required, in hours, for its completion.

Activity	Depends on
A(3)	–
B(5)	–
C(2)	A
D(3)	A
E(3)	B, D
F(5)	C, E
G(1)	C
H(2)	F, G

A precedence table can be used to produce an **activity network**. In the network, activities are represented by arcs and events are represented by vertices.

The vertices are numbered from 0 at the **start vertex** and finishing at the **terminal vertex**.

The direction of the arrows shows the order in which the activities must be completed.

There must only be *one* activity between each pair of events in the network. The notation (i, j) is used to represent the activity between events i and j.

A **dummy activity** is one that has zero duration. A dummy is needed in this network to show that G depends on C whereas F depends on C *and* E.

A dummy is shown with a dotted line. Its direction is important in defining dependency. In this case it shows that F depends on C not that G depends on E.

Earliest event times

The **earliest event time** for vertex i is denoted by e_i and represents the earliest time of arrival at event i with all dependent activities completed. These times are calculated using a **forward scan** from the start vertex to the terminal vertex.

Latest event times

The **latest event time** for vertex i is denoted by l_i and represents the latest time that event i may be left without extending the time for the project. These times are calculated using a **backward scan** from the terminal vertex back to the start vertex.

The **critical path** is the longest path through the network. The activities on this path are the **critical activities**. If any critical activity is delayed then this will increase the time needed to complete the project. The events on the critical path are the **critical events** and for each of these $e_i = l_i$.

It is useful to add the information about earliest and latest times to the network. The critical path is then easily identified.

Notice that B(5) lies between two critical events but it is not a critical activity.

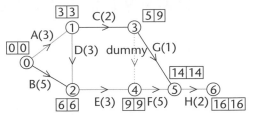

> The total float of any critical activity is always zero.

The **total float of an activity** is the maximum time that the activity may be delayed without affecting the length of the critical path. It is given by:

latest finish time – earliest start time – duration of the activity.

Scheduling

The process of allocating activities to workers for completion, within all of the constraints of the project, is known as **scheduling**.

The information regarding earliest and latest times for each activity is crucial when constructing a schedule. This information may be presented as a table or as a chart.

> Typically, the purpose of scheduling is to determine the number of workers needed to complete the project in a given time, or to determine the minimum time required for a given number of workers to complete the project.

Activity	Duration	Start Earliest	Start Latest	Finish Earliest	Finish Latest	Float
A(0, 1)	3	0	0	3	3	0
B(0, 2)	5	0	1	5	6	1
C(1, 3)	2	3	7	5	9	4
D(1, 2)	3	3	3	6	6	0
E(2, 4)	3	6	6	9	9	0
F(4, 5)	5	9	9	14	14	0
G(3, 5)	1	5	13	6	14	8
H(5, 6)	2	14	14	16	16	0

The critical activities are shown along one line.

> This type of diagram is called a Gantt chart or a cascade chart.

The diagram illustrates the degree of flexibility in starting activities B, C and G. Remember that G cannot be started until C has been completed.

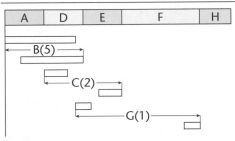

Progress check

1 Determine the critical activities and the length of the critical path for this network.

1 A, D, G and J, 21

5.4 Linear programming

After studying this section you should be able to:

- *formulate a linear programming problem in terms of decision variables*
- *use a graphical method to represent the constraints and solve the problem*

LEARNING SUMMARY

Formulating a linear programming problem

x and *y* are often used for the variables.

Typically, this may be to maximise a profit or minimise a loss.

To formulate a linear programming problem you need to:

- Identify the **variables** in the problem and give each one a label.
- Express the **constraints** of the problem in terms of the variables. You need to include non-negativity constraints such as $x \geqslant 0$, $y \geqslant 0$.
- Express the quantity to be optimised in terms of the variables. The expression produced is called the **objective function**.

Example

A small company produces two types of armchair. The cost of labour and materials for the two types is shown in the table.

	Labour	Materials
Standard	£30	£25
Deluxe	£40	£50

The total spent on labour must not be more than £1150 and the total spent on materials must not be more than £1250. The profit on a standard chair is £70 and the profit on a deluxe chair is £100. How many chairs of each type should be made to maximise the profit?

In this case, the variables are the number of chairs of each type that may be produced. Using x to represent the number of standard chairs and y to represent the number of deluxe chairs, the constraints may be written as:

$$30x + 40y \leqslant 1150 \Rightarrow 3x + 4y \leqslant 115$$
$$25x + 50y \leqslant 1250 \Rightarrow x + 2y \leqslant 50$$

It's a good idea to simplify the constraints where possible.

and $\quad x \geqslant 0, \quad y \geqslant 0.$

Using P to stand for the profit, the problem is to maximise $P = 70x + 100y$.

The graphical method of solution

Each constraint is represented by a region on the graph. It's a good idea to shade out the *unwanted* region for each one. The part that remains unshaded then defines the **feasible region** containing the points that satisfy all of the constraints.

The blue line represents the points where the profit takes a particular value. Moving the line in the direction of the arrow corresponds to increasing the profit. This suggests that the maximum profit occurs at the point X.

X does not represent the solution in this case because both *x* and *y* must be integers.

Solving $3x + 4y = 115$ and $x + 2y = 50$ simultaneously gives X as (15, 17.5).

The nearest points with integer coordinates in the feasible region are (15,17) and (14,18) The profit, given by $P = 70x + 100y$, is greater at (14,18).

The maximum profit is made by producing 14 standard and 18 deluxe chairs.

Progress check

1 In the diagram, the shaded region represents the feasible region for a linear programming problem. All of the boundary lines are included.

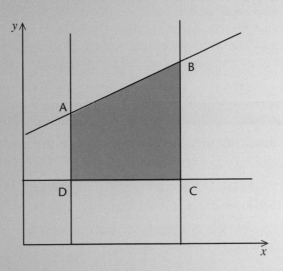

The object is to maximise a profit given by $P = 1.2x + 0.9y$

Which point of the feasible region represents the maximum profit?

1 B (see diagram for explanation)

5.5 Matchings

LEARNING SUMMARY

After studying this section you should be able to:

- *use bipartite graphs to model matchings*
- *understand the conditions for matchings to be maximal or complete*
- *apply the maximum matching algorithm*

Matchings and graphs

A **bipartite graph** is a graph in which the vertices are divided into two sets such that no pair of vertices in the same set is connected by an edge.

In this case, the two sets are {A, B, C} and {p, q, r, s}.

Some vertices in a bipartite graph may not be connected to another vertex.

A **matching** between two sets may be represented by a bipartite graph in which there is at most one edge connecting a pair of vertices.

A **maximal matching** is a matching which has the maximum number of edges. This occurs when every vertex in one of the sets is connected to a vertex in the other set. The bipartite graph shown above represents a maximal matching.

A **complete matching** is a matching in which every vertex is connected to another vertex. This can only occur when the two sets contain the same number of vertices.

The matching improvement algorithm

Figure 1 is a bipartite graph showing the possible connections between two sets. It does not represent a matching because some vertices have more than one connection.

Figure 2 is a bipartite graph representing an initial matching.

An initial matching may be improved by increasing the number of connections. This is the purpose of the **matching improvement algorithm**.

Figure 1

Figure 2

Alternating path

Maximal matching

> In an alternating path, the edges alternate between those that are not in the initial matching and those that are.

Step 1 In the initial matching, start from a vertex that is not connected and look for an **alternating path** to a vertex in the other set that is not connected.

Step 2 Each edge on the alternating path, not included in the initial matching, is now included and each edge originally included is removed.

Step 3 Repeat steps 1 and 2 using the latest matching in place of the initial matching until no further alternating paths can be found.

Progress check

1 Starting with an initial matching in which 1 is connected to R and 2 is connected to P, use the improvement algorithm to establish a complete matching for this bipartite graph.

1 1–P, 2–S, 3–R, 4–Q.

Sample questions and model answers

1

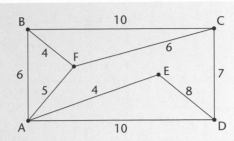

The diagram represents the system of pathways that a security guard must patrol during his course of duty. The weights on the edges represent the time taken, in minutes, to patrol each pathway.

Find the minimum time required to patrol every pathway at least once and give a possible route.

You need to recognise that this question requires the use of the route inspection algorithm.

Step 1 of the algorithm is to list the odd vertices.

Vertex	Order
A	4
B	3
C	3
D	3
E	2
F	3

The odd vertices are B, C, D and F.

The sets of pairs of odd vertices are: {BC, DF}
{BD, CF}
{BF, CD}.

Step 2 of the algorithm is to find the sets of pairs of odd vertices. Each odd vertex appears once in each set.

For the set {BC, DF}

BC = 10. The shortest route from D to F is DC + CF = 13.

There is a an edge connecting B and C.

Total length of the extra paths for this set is 10 + 13 = 23.

Step 3 of the algorithm is to find the extra length introduced for each set.

For the set {BD, CF}

The shortest route from B to D is BA + AD = 16. CF = 6.

Total length of the extra paths for this set is 16 + 6 = 22.

For the set {BF, CD}

There is an edge connecting C and F.

BF = 4 and CD = 7.

Total length of the extra paths for this set is 4 + 7 = 11.

The minimum total corresponds to the set {BF, CD} which means that the edges BF and CD will be repeated.

There are many possible routes.

A possible route is ABFCDEAFBCDA.

Step 4 of the algorithm is to keep going until all the sets have been considered.

The time taken will be 60 + 11 = 71 minutes.

Sample questions and model answers *(continued)*

2

The diagram is an activity network for a project.

The figures in brackets represent the time in days to complete each activity.

The start vertex is S and the terminal vertex is T.

(a) Label each vertex with the earliest and latest time for the event. Give the length of the critical path.

(b) State which events are critical and give the critical path.

> Use a forward scan to find the earliest event times.

> Use a backward scan to find the latest event times.

(a) The length of the critical path is 51 days.

(b) The critical events are S, 1, 2, 3, 4, 6 and T.
   The critical path is A, C, D, E, G and J.

> The length of the critical path is given by the earliest (and latest) event time at the terminal vertex.

Practice examination questions

1   The table shows the amount of memory taken up by some files on a computer system.

File	A	B	C	D	E	F	G
Memory (Kb)	580	468	610	532	840	590	900

The files are to be transferred onto disks with a capacity of 1.4 Mb (1000 Kb = 1 Mb).

(a) Show that a minimum of four disks is required.

(b) Use the first-fit algorithm to allocate the files to disks.

(c) Use the first-fit decreasing algorithm to show how all of the files may be stored using four disks.

2   A new theme park has its major attractions at A, B, C, D, E, F and G as shown in the network below. The edges of the network show the *possible* routes of paths connecting the attractions.

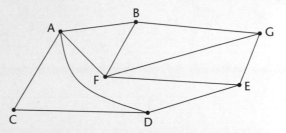

The cost of laying each of these paths, in hundreds of pounds, is given in the table:

	A	B	C	D	E	F	G
A	–	50	40	61	–	38	–
B	50	–	–	–	–	47	54
C	40	–	–	35	–	–	–
D	61	–	35	–	42	–	–
E	–	–	–	42	–	25	10
F	38	47	–	–	25	–	63
G	–	54	–	–	10	63	–

The paths that are actually laid must form a connected network.
Use Prim's algorithm, starting by deleting row A, to find the minimum cost of laying the necessary paths.

Practice examination questions (continued)

3

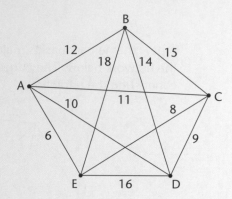

(a) Use Dijkstra's algorithm to find the shortest distance from G to D.

(b) Describe the route taken.

4

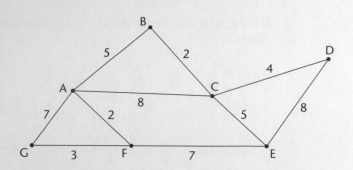

(a) Use Kruskal's algorithm to find the length of the minimum spanning tree for this network.

State the order in which you include each of the edges.

(b) Find an upper bound for the travelling salesman problem using the minimum spanning tree with shortcuts.

5

(i)   (ii)   (iii)

List the order of the vertices for each graph.

Practice examination questions *(continued)*

6 A distribution manager has the task of transporting pallets of frozen food between two storage centres. The maximum number of pallets is to be delivered within a daily budget of £2000. Three types of van are available and each van can only do the trip once per day.

The details of what the vans can carry and the daily costs are shown in the table.

At most 10 drivers may be used in one day.

Van type	No. of pallets	Cost/day
Class A	4	£120
Class B	9	£300
Class C	12	£400

Let a represent the number of Class A vans used, b the number of Class B vans used, c the number of Class C vans used and P the total number of pallets transported in a day.

(a) Write an expression for the objective function.

(b) Formulate the task as a linear programming problem.

(c) Describe any special condition that the solution must satisfy.

7 The table shows the names of five employees of a company and the days when they are each available to work a late shift. The days that have been highlighted represent a first attempt at matching the employees to the available days to make a roster.

(a) Use a bipartite graph to represent the relationship between employees and days available.

(b) Use a second bipartite graph to represent an initial matching based on the days highlighted.

(c) Implement a matching improvement algorithm to obtain a complete matching. Indicate the alternating path used.

Name	Days available
Abbie	**Mon**, Tue, Fri
Ben	Mon, **Wed**, Thu
Colin	Mon, **Fri**
Dave	Fri, Wed
Emma	**Thu**, Fri

Practice examination answers

Core 1

1 $49^{-\frac{1}{2}} = \dfrac{1}{49^{\frac{1}{2}}} = \dfrac{1}{\sqrt{49}} = \dfrac{1}{7}.$

2 $8^{x-3} = 4^{x+1} \Rightarrow (2^3)^{x-3} = (2^2)^{x+1}$

$\Rightarrow 2^{3(x-3)} = 2^{2(x+1)}$

$\Rightarrow 3(x-3) = 2(x+1)$

$\Rightarrow x = 11.$

3   (a) $x^2 - 4x + 1 = (x-2)^2 - 3.$

(b) $(x-2)^2 - 3 = 0$

$\Rightarrow x - 2 = \pm\sqrt{3}$

$\Rightarrow x = 2 \pm \sqrt{3}.$

(c) $(2, -3).$

(d)

4   (a)   (b)

5   The coordinates at the points of intersection are given by the simultaneous solution of:

$$y = 2x - 2 \qquad\qquad [1]$$

$$y = x^2 - x - 6 \qquad\qquad [2]$$

Eliminating y gives

$$x^2 - x - 6 = 2x - 2$$

$\Rightarrow \qquad x^2 - 3x - 4 = 0$

$\Rightarrow \qquad (x+1)(x-4) = 0$

$\Rightarrow \qquad x = -1 \text{ or } x = 4.$

When $x = -1$, $y = -4.$

When $x = 4$, $y = 6.$

The points of intersection are A$(-1, -4)$ and B$(4, 6)$.

6 $3x^2 - 8x - 7 < 2x^2 - 3x - 11$

$\Rightarrow x^2 - 5x + 4 < 0$

$\Rightarrow (x-1)(x-4) < 0$

$\Rightarrow 1 < x < 4.$

7   (a)   Gradient of AB $= 6/5.$

Using $y - y_1 = m(x - x_1)$ gives

$$y - 1 = 6/5(x - 4)$$

$\Rightarrow \qquad 5y - 5 = 6x - 24$

$\Rightarrow \qquad 6x - 5y - 19 = 0.$

(b)   The mid point of A and B is $(1.5, -2)$.

The gradient of the line l is $-5/6$.

Using $y - y_1 = m(x - x_1)$ again gives

$$y + 2 = -5/6(x - 1.5)$$

$\Rightarrow \qquad 6y + 12 = -5x + 7.5$

$\Rightarrow \qquad 12y + 24 = -10x + 15$

$\Rightarrow \qquad 10x + 12y + 9 = 0.$

(c)   At C, $x = 0 \Rightarrow y = -0.75$ so the coordinates of C are $(0, -0.75)$.

Core 1 (continued)

8 (a) The equation of the line m is $3x + 2y = 6$

$\Rightarrow \quad y = -3/2x + 3$

$\Rightarrow \quad$ The gradient of the line m is $-3/2$

$\Rightarrow \quad$ The gradient of the line l is $2/3$.

Using $y - y_1 = m(x - x_1)$ gives

$y - 4 = 2/3(x - 5)$

$\Rightarrow \quad 3y - 12 = 2x - 10$

$\Rightarrow \quad 2x - 3y + 2 = 0$ which is the equation of line l.

(b) At A, $y = 0 \Rightarrow x = -1$.

B lies on the line with equation $3x + 2y = 6$.

At B, $y = 0 \Rightarrow x = 2$.

So the distance AB is given by $2 - (-1) = 3$.

9 10th term $= a + 9d = 74$. [1]

Sum of first 20 terms $= 10(2a + 19d) = 1510$

$\Rightarrow \qquad\qquad 2a + 19d = 151$ [2]

$2 \times$ [1] gives $\qquad 2a + 18d = 148$ [3]

[2] – [3] gives $\qquad\qquad d = 3$

Subs for d in [1] gives $a + 27 = 74 \Rightarrow a = 47$.

First term $= 47$, common difference $= 3$.

10 (a) $x + y = 7$

(b) At P and Q $\qquad x + y = 7$ [1]

and $\qquad\qquad y = (x - 1)^2 + 4$ [2]

Substituting for y in [2] gives

$7 - x = x^2 - 2x + 5$

$\Rightarrow x^2 - x - 2 = 0$

$\Rightarrow (x - 2)(x + 1) = 0$

$\Rightarrow x = 2$ or $x = -1$.

$x = 2 \Rightarrow y = 5$ and $x = -1 \Rightarrow y = 8$.

P is the point $(-1, 8)$ and Q is the point $(2, 5)$.

Core 2

1 (a) First term $a = 3$, common ratio $r = 2$.

The 8th term is $ar^7 = 3 \times 2^7 = 384$.

(b) First term $a = 25$, common difference $d = 1.2$

number of terms $n = 50$.

Using $S_n = \dfrac{n}{2}(2a + (n - 1)d)$ gives

Sum $= \frac{50}{2}(2 \times 25 + 49 \times 1.2) = 2720$.

2 (a) By the factor theorem $f(-2) = 0$

$\Rightarrow (-2)^3 - 4(-2)^2 - 3(-2) + k = 0$

$\Rightarrow -8 - 16 + 6 + k = 0$

$\Rightarrow k = 18$.

(b) $f(x) = x^3 - 4x^2 - 3x + 18$

$\equiv (x + 2)(Ax^2 + Bx + C)$.

Comparing coefficients of x^3 gives A = 1.

Comparing the constant terms gives C = 9.

When $x = -1$

$-1 - 4 + 3 + 18 = A - B + C$

$\Rightarrow 16 = 1 - B + 9 \Rightarrow B = -6$.

$f(x) = (x + 2)(x^2 - 6x + 9)$

$= (x + 2)(x - 3)^2$.

(c) From the sketch, $f(x) \geqslant 0 \Rightarrow x \geqslant -2$.

(d) By the remainder theorem, the remainder is given by $f(-1) = 16$.

Core 2 (continued)

3 (a) $3\log_a x - 2\log_a y + \log_a(x+1)$
$= \log_a x^3 - \log_a y^2 + \log_a(x+1)$
$= \log_a\left(\dfrac{x^3(x+1)}{y^2}\right).$

 (b) $5^x = 100$
$\Rightarrow \log_{10} 5^x = \log_{10} 100$
$x\log_{10} 5 = \log_{10} 100$
$x = \dfrac{\log_{10} 100}{\log_{10} 5} = 2.861$ to 3 d.p.

4 $\cos^{-1}(0.4) = 66.42\ldots°$
$0° \leqslant x \leqslant 360° \Rightarrow 30° \leqslant 2x + 30 \leqslant 750°$
Sketch the graph of $y = \cos X$, $30° \leqslant X \leqslant 750°$,
and the graph of $y = 0.4$

From the sketch, the values of X (i.e $2x + 30$) are:
$66.42\ldots°$, $360 - 66.42\ldots°$, $360 + 66.42\ldots°$,
$720 - 66.42\ldots°$,
This gives
$x = 18.2°$ or $x = 131.8°$ or $x = 198.2°$ or $x = 311.8°$.

5 (a) $\cos x + 3\sin x \tan x - 2 = 0$
$\Rightarrow \cos^2 x + 3\sin^2 x - 2\cos x = 0$
$\Rightarrow \cos^2 x + 3(1 - \cos^2 x) - 2\cos x = 0$
$\Rightarrow 3 - 2\cos^2 x - 2\cos x = 0$
$\Rightarrow 2\cos^2 x + 2\cos x - 3 = 0$

 (b) $\cos x = \dfrac{-2 \pm \sqrt{4 + 24}}{4}$
$\Rightarrow \cos x = 0.8228\ldots$ or $\cos x = -1.8228\ldots$
                          (No solutions)
$\Rightarrow x = 34.6°$ or $x = 325.4°$

6 (a) $f'(x) = 2x - x^{-2}$
$f'(2) = 4 - \frac{1}{4} = 3\frac{3}{4}$
 (b) $f''(x) = 2 + 2x^{-3}$
$f''(-1) = 2 - 2 = 0.$

7 Area $= \displaystyle\int_1^4 3x^{\frac{1}{2}}\, dx$
$= [2x^{\frac{3}{2}}]_1^4 = 2(8 - 1)$
$= 14.$

8 Area sector = area triangle + area segment
$\Rightarrow \dfrac{1}{2}\theta = \dfrac{1}{2}\sin\theta + \dfrac{\pi}{5}$
$\Rightarrow \theta = \sin\theta + \dfrac{2\pi}{5}.$

9 Area $= \displaystyle\int_2^3 \sqrt{x^2 - 3}\, dx.$
$d = \dfrac{3-2}{5} = 0.2$ Taking $f(x) = \sqrt{x^2 - 3}$ gives
$A = 0.1(f(2) + 2(f(2.2) + f(2.4) + f(2.6) + f(2.8)) + f(3))$
$A = 0.1(1 + 2(1.3565 + 1.6613 + 1.9591 + 2.2) + 2.4495)$
$A = 1.78$ to 2 d.p.

10 (a) $2x^{\frac{3}{2}}\left(1 - \dfrac{x}{5}\right)$

 (b) $(0, 0)$, $(5, 0)$

 (c) $\dfrac{dy}{dx} = 3x^{\frac{1}{2}} - x^{\frac{3}{2}}$

 (d) $\dfrac{dy}{dx} = 0 \Rightarrow x^{\frac{1}{2}}(3 - x) = 0$
$\Rightarrow x = 0$ or $x = 3$
Since $x > 0$, $x = 3$
when $x = 3$, $y = 4.157$ (4 s.f.)
$\dfrac{d^2 y}{dx^2} = \frac{3}{2}x^{-\frac{1}{2}} - \frac{3}{2}x^{\frac{1}{2}}$
when $x = 3$, $\dfrac{d^2 y}{dx^2} = \frac{3}{2}(3^{-\frac{1}{2}} - 3^{\frac{1}{2}}) < 0$
So the turning point at $(3, 4.157)$ is a maximum.

11 (a) $y = 2x^3 - 15x^2 - 36x + 10$
$\dfrac{dy}{dx} = 6x^2 - 30x - 36$
At a stationary point
$\dfrac{dy}{dx} = 0 \Rightarrow 6x^2 - 30x - 36 = 0$
$\Rightarrow x^2 - 5x - 6 = 0$
$\Rightarrow (x + 1)(x - 6) = 0$
$\Rightarrow x = -1$ or $x = 6$.
When $x = -1$, $y = 29$.
When $x = 6$, $y = -314$.
The stationary points are $(-1, 29)$ and $(6, -314)$.

 (b) $\dfrac{d^2 y}{dx^2} = 12x - 30$
When $x = -1$, $\dfrac{d^2 y}{dx^2} = -42 < 0$ maximum.
When $x = 6$, $\dfrac{d^2 y}{dx^2} = 42 > 0$ minimum.

 (c) The function is decreasing at all of the points
between the turning points i.e. when $-1 < x < 6$.

Mechanics 1

1 (a)
$$u = 10$$
$$a = -4$$
Using $v = u + at$ $v = 10 - 4 \times 5 = -10.$
The speed of the particle when $t = 5$ is 10 m s^{-1}.

(b)
$$v = 0 \Rightarrow 10 - 4t = 0$$
$$\Rightarrow t = 2.5$$
Using $s = ut + \frac{1}{2}at^2$
when $t = 2.5$ $s = 10 \times 2.5 - 2 \times 6.25$
$$s = 12.5$$
when $t = 4$ $s = 10 \times 4 - 2 \times 16$
$$s = 8.$$
The total distance travelled between $t = 0$ and $t = 4$
is 12.5 m + 4.5 m = 17 m.

2 (a) The acceleration is positive between $t = 0$ and
$t = 25$ so the greatest velocity occurs when $t = 25$.

(b)

The area under the graph represents the increase
in velocity between $t = 0$ and $t = 25$.
Area $= \frac{1}{2} \times 10(5 + 8) + \frac{1}{2} \times 15 \times 8 = 125.$
The increase in velocity is 125 m s^{-1}.
The initial velocity is 20 m s^{-1}.
The maximum value of the velocity is 145 m s^{-1}.

3 (a) The acceleration is constant so the constant
acceleration formulae may be used.
$$\mathbf{u} = 2\mathbf{i} + \mathbf{j}$$
$$\mathbf{a} = \mathbf{i} + 3\mathbf{k}$$
$$t = 4.$$
Using $\mathbf{s} = \mathbf{u}t + \frac{1}{2}\mathbf{a}t^2$
$$\mathbf{s} = (2\mathbf{i} + \mathbf{j})4 + (\mathbf{i} + 3\mathbf{k})8$$
$$\mathbf{s} = 16\mathbf{i} + 4\mathbf{j} + 24\mathbf{k}.$$
This is the displacement from P to Q
so $\overrightarrow{PQ} = 16\mathbf{i} + 4\mathbf{j} + 24\mathbf{k}.$

(b) $\overrightarrow{OQ} = \overrightarrow{OP} + \overrightarrow{PQ}$
$$= -5\mathbf{i} + 11\mathbf{j} + \mathbf{k} + 16\mathbf{i} + 4\mathbf{j} + 24\mathbf{k}$$
$$= 11\mathbf{i} + 15\mathbf{j} + 25\mathbf{k}.$$

(c) Average speed $= \dfrac{\text{distance}}{\text{time}} = \dfrac{|\overrightarrow{PQ}|}{4}$ m s^{-1}
$$|\overrightarrow{PQ}| = \sqrt{16^2 + 4^2 + 24^2} = 29.12$$
Average speed = 7.28 m s^{-1} to 2 d.p.

4

(a) Resolving vertically
$$R + 50 \sin 40° - 120 = 0$$
$$\Rightarrow R = 120 - 50 \sin 40°$$
$$\Rightarrow R = 87.86 \ldots.$$
The normal reaction is 87.9 N to 1 d.p.

(b) Resolving horizontally
$$F - 50 \cos 40° = 0$$
$$\Rightarrow F = 38.30 \ldots.$$
For limiting equilibrium
$$F = \mu R$$
giving $38.30 \ldots = \mu \times 87.86 \ldots$
$$\mu = 0.4359 \ldots.$$
The coefficient of friction is 0.436 to 3 d.p.

5

*The equations are simpler to deal with by resolving
vertically and horizontally, in this case, instead of parallel
and normal to the plane.*

The minimum value of P corresponds to the case where
friction is limiting, as shown in the diagram.

Resolving vertically
$$R \cos 30° + 0.3 R \sin 30° = 6 \times 9.8$$
$$R = \frac{6 \times 9.8}{\cos 30° + 0.3 \sin 30°}$$
$$R = 57.87 \ldots.$$

Resolving horizontally
$$P + 0.3R \cos 30° - R \sin 30° = 0$$
$$P = R \sin 30° - 0.3R \cos 30°$$
$$P = 13.90 \ldots.$$

The minimum value of P is 13.9 to 1 d.p.

6

(a) The equations of motion for the particles are:

For A $\quad T - 19.6 = 2a$

For B $\quad 29.4 - T = 3a$.

Adding gives $9.8 = 5a \Rightarrow a = 1.96$

The acceleration of the system is 1.96 m s^{-2}

(b) $T - 19.6 = 3.92 \Rightarrow T = 23.52$

The tension in the string is 23.52 N.

(c) For B $\quad u = 0$

$a = 1.96,$

$s = 1.$

Using $\quad s = ut + 1/2at^2$

$1 = 0 + 0.98t^2 \; (t > 0)$

$\Rightarrow t = 1.01 \dots .$

Particle B hits the ground after 1.01 s to 2 d.p.

(d) using $\quad v = u + at$

$v = 0 + 1.96 \times 1.01 \dots = 1.979 \dots .$

Particle B hits the ground with speed 1.98 m s^{-1} to 2 d.p.

(e) Particle A rises to height 2 m then the string becomes slack and it behaves as a projectile.

Using $\qquad v^2 = u^2 + 2as$

at max height $\qquad 0 = 1.979^2 - 19.6s$

$\Rightarrow s = 0.2$

The max height reached by particle A is 2.2 m.

(f) In the extreme situation where the pulley does not move, each particle would be held in equilibrium and so the tension on each side would match the weight of the corresponding particle.

In the given situation, where the pulley is not light or frictionless, it cannot be assumed that the tension on each side of the pulley is the same. This introduces an extra unknown value into the equations. For example, the tension on one side might be labelled T_1 and on the other T_2.

7 (a) Before impact

Momentum before = momentum after

so $\qquad 10mu = 2mv + 2mv$

$\Rightarrow 10u = 4v$

$\Rightarrow v = 2.5u$

(b) Impulse = change in momentum

$= 5mu - 0$

$= 5mu$

Statistics 1

1 (a) $\Sigma p(x) = 1$

$\Rightarrow \dfrac{1}{k} + \dfrac{4}{k} + \dfrac{9}{k} = 1$

$\Rightarrow \dfrac{14}{k} = 1$

$k = 14.$

(b) Mean $= E(X) = \Sigma x p(x)$

$= 1 \times \dfrac{1}{14} + 2 \times \dfrac{4}{14} + 3 \times \dfrac{9}{14} = \dfrac{36}{14}$

$= \dfrac{18}{7}.$

$\mathrm{Var}(X) = E(X^2) - (E(X))^2$

$= x^2 p(x) - \left(\dfrac{18}{7}\right)^2$

$= \dfrac{1}{14} + 4 \times \dfrac{4}{14} + 9 \times \dfrac{9}{14} - \left(\dfrac{18}{7}\right)^2$

$= \dfrac{19}{49}.$

(c) Mean of $3X - 5$ is $3 \times \dfrac{18}{7} - 5 = \dfrac{19}{7}.$

Variance of $3X - 5$ is $9 \times \dfrac{19}{49} = \dfrac{171}{49}.$

2 (a) $P(A \cup B) = P(A) + P(B) - P(A \cap B)$

$0.72 = 0.3 + 0.6 - P(A \cap B)$
$\Rightarrow P(A \cap B) = 0.18$
As $P(A \cap B) \neq 0$
A and B are not mutually exclusive.

(b) $P(A) \times P(B) = 0.3 \times 0.6 = 0.18$
As $P(A \cap B) = P(A) \times P(B)$,
A and B are independent.

(c) $P(A \cup B)' = 1 - P(A \cup B) = 0.28$

(d) $P(A' \cup B') = P(A') + P(B') - P(A' \cap B')$

$= 0.7 + 0.4 - 0.7 \times 0.4$

$= 0.82$

3 (a) (i) $P(RRR) = \dfrac{26}{52} \times \dfrac{25}{51} \times \dfrac{24}{50} = \dfrac{2}{17}$

(ii) $P(\text{same colour}) = P(RRR) + P(BBB)$

$= \dfrac{2}{17} + \dfrac{2}{17} = \dfrac{4}{17}.$

(iii) $P(\text{at least one red}) = 1 - P(BBB)$

$= 1 - \dfrac{2}{17} = \dfrac{15}{17}.$

(b) Using A to represent getting an ace and D to represent getting a diamond:

$P(A|D') = \dfrac{P(A \cap D')}{P(D')} = \dfrac{\frac{3}{52}}{\frac{3}{4}} = \dfrac{3}{52} \times \dfrac{4}{3}$

$= \dfrac{1}{13}.$

4 Using X to represent the lifetime of a battery in hours:

(a) $P(X > 8.5) = P\left(Z > \dfrac{8.5 - 8}{\frac{25}{60}}\right)$

$= P(Z > 1.2)$

$= 1 - \Phi(1.2)$

$= 1 - 0.8849 \ldots$

$= 0.1151 \ldots .$

So the probability that a battery selected at random will last more than $8\frac{1}{2}$ hours is 0.115 to 3 d.p.

(b) $P(X < 7) = P\left(Z < \dfrac{7 - 8}{\frac{25}{60}}\right)$

$= P(Z < -2.4)$

$= \Phi(-2.4)$

$= 1 - \Phi(2.4)$

$= 1 - 0.9918$

$= 0.0082$

The percentage of batteries that will last for less than 7 hours is 0.82%

Statistics 1 (continued)

5 (a) $r = \dfrac{S_{xy}}{\sqrt{S_{xx}S_{yy}}}$

$$= \frac{28785 - \dfrac{567 \times 484}{10}}{\sqrt{\left\{ \left(34216 - \dfrac{576^2}{10}\right)\left(24450 - \dfrac{484^2}{10}\right) \right\}}}$$

$= 0.879$ to 3 d.p.

(b) $b = \dfrac{S_{xy}}{S_{xx}} = \dfrac{28785 - \dfrac{576 \times 484}{10}}{34216 - \dfrac{576^2}{10}}$

$= 0.87307... = 0.873$ (3 d.p.)

$a = \dfrac{484}{10} - 0.873... \times \dfrac{576}{10} = -1.88906...$

The equation is $y = -1.889 + 0.873x$.

(c) $\hat{y} = -1.889 + 0.873 \times 65 = 55$ (2 s.f.)

If the assumption that physics scores depend on maths scores is correct, the estimate should be reliable, since $x = 65$ is within the range of data and the high value of r suggests a strong linear correlation between the variables.

Decision Mathematics 1

1 (a) Total memory to transfer = 4520 Kb = 4.52 Mb

$\dfrac{4.52}{1.4} = 3.228...$

Three disks do not have sufficient capacity to hold all of the files. A minimum of four disks is required.

(b)

(c) Writing the memory sizes in order gives
900, 840, 610, 590, 580, 532, 468.

2 Applying the matrix form of Prim's algorithm gives:

	1	7	5	6	3	2	4
	A	B	C	D	E	F	G
A	–	50	40	61	–	38	–
B	50	–	–	–	–	(47)	54
C	(40)	–	–	35	–	–	–
D	61	–	(35)	–	42	–	–
E	–	–	–	42	–	(25)	10
F	(38)	47	–	–	25	–	63
G	–	54	–	–	(10)	63	–

The minimum cost is found by adding the circled figures and multiplying by £100.

Minimum cost = £19 500.

3

The shortest distance from G to D is 16.

(a) The shortest distance from G to D is 16.

(b) The route is G, F, A, B, C, D.

Decision Mathematics 1 *(continued)*

4

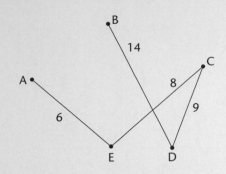

(a) The order in which the edges were connected is:

AE, EC, CD, AB (not AD or AC to avoid cycles).

The length of the minimum spanning tree is 37.

(b) Using the short-cut from B to A gives an upper bound for the TSP as $37 + 12 = 49$.

5

Vertices					
Graph (i)	A	B	C	D	E
Order	4	2	4	3	3

Vertices						
Graph (ii)	A	B	C	D	E	F
Order	4	2	4	4	4	2

Vertices				
Graph (iii)	A	C	D	E
Order	3	3	3	3

6 (a) The objective function is $4a + 9b + 12c$.

(b) The formulation of the task as a linear programming problem is:

Maximise: $P = 4a + 9b + 12c$

Subject to: $6a + 15b + 20c \leqslant 100$

$\qquad\qquad a + b + c \leqslant 10$

and $\qquad a \geqslant 0, b \geqslant 0, c \geqslant 0$.

(c) In the solution a, b and c must be integers.

7 (a)

(b)

(c)

The alternating path is:

D – Wed – B – Mon – A – Tue.

The new matching is:

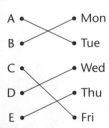

Notes

Notes

Index

Revise
A2

Edexcel
Mathematics

Peter Sherran & Janet Crawshaw

Contents

Chapter 3 Mechanics 2

Chapter 4 Decision Mathematics 2

Specification information

Edexcel Mathematics

MODULE	SPECIFICATION TOPIC	CHAPTER REFERENCE	STUDIED IN CLASS	REVISED	PRACTICE QUESTIONS
Core 3 (C3)	Algebra and functions	1.1, 1.2			
	Trigonometry	1.4			
	Exponentials and logarithms	1.2			
	Differentiation	1.5			
	Numerical methods	1.7			
Core 4 (C4)	Algebra and functions	1.1			
	Coordinate geometry in the (x, y) plane	1.3, 1.6			
	Sequences and series	1.1			
	Differentiation	1.5			
	Integration	1.6, 1.7			
	Vectors	1.8			
Statistics 2 (S2)	Binomial and Poisson distributions	2.1			
	Continuous random variables	2.2			
	Continuous distributions	2.2			
	Hypothesis tests	2.3, 2.4			
Mechanics 2 (M2)	Kinematics	3.1			
	Centres of mass	3.2			
	Work and energy	3.3			
	Collisions	3.4			
	Statics of rigid bodies	3.5			
Decision 2 (D2)	Transportation problems	4.2			
	Allocation problems	4.3			
	The travelling salesman problem	4.5			
	Further linear programming	4.1			
	Game theory	4.4			
	Flows in networks	4.5			
	Dynamic programming	4.6			

Examination analysis

The Advanced Level Mathematics GCE consists of AS (50%) and A2 (50%). AS consists of C1 + C2 + one of M1, S1, D1. A2 consists of C3 + C4 + one of M1, S1, D1, M2, S2, D2 (following any dependency rules).

C3	A2	Scientific / graphics calculator	1 hr 30 min exam	$16\frac{2}{3}$% of A Level
C4	A2	Scientific / graphics calculator	1 hr 30 min exam	$16\frac{2}{3}$% of A Level
M1/M2	AS/A2	Scientific / graphics calculator	1 hr 30 min exam	$16\frac{2}{3}$% of A Level
S1/S2	AS/A2	Scientific / graphics calculator	1 hr 30 min exam	$16\frac{2}{3}$% of A Level
D1/D2	AS/A2	Scientific / graphics calculator	1 hr 30 min exam	$16\frac{2}{3}$% of A Level

Your AS/A2 Level Mathematics course

AS and A2

The Edexcel Mathematics A Level course is made up of six modules. Students first study the AS (Advanced Subsidiary) course, which has three modules. Some will then go on to study the second part of the A Level course, a further three modules called A2. Advanced Subsidiary is assessed at the standard expected halfway through an A Level course: i.e., between GCSE and Advanced GCE. This means that the AS and A2 courses are designed so that difficulty steadily increases:

- AS Mathematics builds from GCSE Mathematics
- A2 Mathematics builds from AS Mathematics.

How will you be tested?

Assessment units

For AS Mathematics, you will be tested by three assessment units. For the full A Level in Mathematics, you will take a further three units. AS Mathematics forms 50% of the assessment weighting and A2 Mathematics forms the other 50% for the full A Level.

Most units can be taken in either January or June, but some A2 units must be taken at the end of the course. There is a lot of flexibility about when exams can be taken and the diagram below shows just some of the ways that the assessment units may be taken for AS and A Level Mathematics.

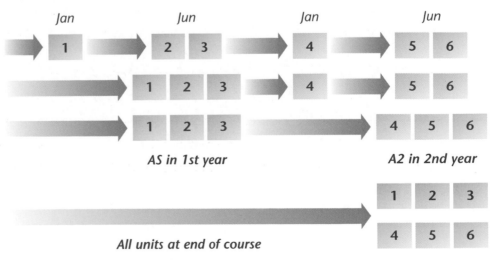

If you are disappointed with a module result, you can resit the module. There is no restriction on the number of times a module may be attempted. The best available results for each module will count towards the final grade.

A2 progression

After having studied AS Mathematics, to continue studying Mathematics to A Level you will need to take three further units of Mathematics, at least two of which are at A2 standard.

Synoptic assessment

Synoptic assessment in Mathematics addresses candidates' understanding of the connections between different elements of the subject. It involves the explicit drawing together of knowledge, understanding and skills learned in different parts of the A Level course, through using and applying methods developed earlier to solving problems. Making and understanding connections in this way is intrinsic to learning mathematics.

For A2 modules you will be asked to demonstrate knowledge of modules studied at AS Level. The contents of C1 and C2 are assumed for C3 and the contents of C1, C2 and C3 are assumed for C4. Similarly in the application modules, a knowledge of S1 is assumed for S2, a knowledge of M1 is assumed for M2 and a knowledge of D1 is assumed for D2.

Key skills

It is important that you develop your key skills throughout your AS and A2 courses. These are important skills that you need whatever you do beyond AS and A Levels. To gain the key skills qualification, which is equivalent to an AS Level, you will need to collect evidence together in a 'portfolio' to show that you have attained Level 3 in Communication, Application of number and Information technology. You will also need to take a formal test in each key skill. You will have many opportunities during AS and A2 Mathematics to develop your key skills.

It is a worthwhile qualification, as it demonstrates your ability to put your ideas across to other people, collect data and use up-to-date technology in your work.

What skills will I need?

For A Level Mathematics (AS and A2), you will be tested by assessment objectives: these are the skills and abilities that you should have acquired by studying the course. The assessment objectives for A Level Mathematics are shown below.

Candidates should be able to:

- recall, select and use their knowledge of mathematical facts, concepts and techniques in a variety of contexts

- construct rigorous mathematical arguments and proofs through use of precise statements, logical deduction and inference and by the manipulation of mathematical expressions, including the construction of extended arguments for handling substantial problems presented in unstructured form

- recall, select and use their knowledge of standard mathematical models to represent situations in the real world; recognise and understand given representations involving standard models; present and interpret results from such models in terms of the original situation, including discussion of the assumptions made and refinement of such models

- comprehend translations of common realistic contexts into mathematics; use the results of calculations to make predictions, or comment on the context; and, where appropriate, read critically and comprehend longer mathematical arguments or examples of applications

- use contemporary calculator technology and other permitted resources (such as formulae booklets or statistical tables*) accurately and efficiently; understand when not to use such technology, and its limitations; give answers to appropriate accuracy.

* A copy of the formulae booklet and statistical tables can be found on the A Level Mathematics section of the Edexcel website, www.edexcel.com

Exam technique

Examiners use certain words in their instructions to let you know what they are expecting in your answer. Make sure that you know what they mean so that you can give the right response.

Write down, state

You can write your answer without having to show how it was obtained. There is nothing to prevent you doing some working if it helps you, but if you are doing a lot then you might have missed the point.

Calculate, find, determine, show, solve

Make sure that you show enough working to justify the final answer or conclusion. Marks will be available for showing a correct method.

Deduce, hence

This means that you are expected to use the given result to establish something new. You must show all of the steps in your working.

Draw

This is used to tell you to plot an accurate graph using graph paper. Take note of any instructions about the scale that must be used. You may need to read values from your graph.

Sketch

If the instruction is to sketch a graph then you don't need to plot the points but you will be expected to show its general shape and its relationship with the axes. Indicate the positions of any turning points and take particular care with any asymptotes.

Find the exact value

This instruction is usually given when the final answer involves an irrational value such as a logarithm, e, π or a surd. You will need to demonstrate that you can manipulate these quantities so don't just key everything into your calculator or you will lose marks.

If a question requires the final answer to be given to a specific level of accuracy then make sure that you do this or you might needlessly lose marks.

Some dos and don'ts

Dos

Do read the question

- Make sure that you are clear about what you are expected to do. Look for some structure in the question that may help you take the right approach.
- Read the question *again* after you have answered it as a quick check that your answer is in the expected form.

Do use diagrams

- In some questions, particularly in mechanics, a clearly labelled diagram is essential. Use a diagram whenever it may help you understand or represent the problem that you are trying to solve.

Do take care with notation

- Write clearly and use the notation accurately. Use brackets when they are required.
- Even if your final answer is wrong, you may earn some marks for a correct expression in your working.

Do learn relevant formulae

- Each module specification contains a list of formulae that *will not be given* in the examination. Make sure that you learn these formulae and practise using them.
- Some formulae *will be given* in the Examination Formulae Booklet. Make sure you know what they are and where to find them. This will help you refer to them quickly in the examination so that you don't waste time.

Do avoid silly answers

- Check that your final answer is sensible within the context of the question.

Do make good use of time

- Choose the order in which you answer the questions carefully. Do the ones that are easiest for you first.
- Set yourself a time limit for a question depending on the number of marks available.
- Be prepared to leave a difficult part of a question and return to it later if there is time.
- Towards the end of the exam make sure that you pick up all of the easy marks in any questions that you haven't got time to answer fully.

Don'ts

Don't work with rounded values

- There may be several stages in a solution that produce numerical values. Rounding errors from earlier stages may distort your final answer. One way to avoid this is to make use of your calculator memories to store values that you will need again.

Don't cross out work that may be partly correct

- It's tempting to cross out something that hasn't worked out as it should. Avoid this unless you have time to replace it with something better.

Don't write out the question

- This wastes time. The marks are for your solution!

What grade do you want?

Everyone should be able to improve their grades but you will only manage this with a lot of hard work and determination. The details given below describe a level of performance typical of candidates achieving grades A, C or E. You should find it useful to read and compare the expectations for the different levels and to give some thought to the areas where you need to improve most.

Grade A candidates

- Recall or recognise almost all the mathematical facts, concepts and techniques that are needed, and select appropriate ones to use in a variety of contexts.
- Manipulate mathematical expressions and use graphs, sketches and diagrams, all with high accuracy and skill.
- Use mathematical language correctly and proceed logically and rigorously through extended arguments or proofs.
- When confronted with unstructured problems they can often devise and implement an effective solution strategy.
- If errors are made in their calculations or logic, these are sometimes noticed and corrected.
- Recall or recognise almost all the standard models that are needed, and select appropriate ones to represent a wide variety of situations in the real world.
- Correctly refer results from calculations using the model to the original situation; they give sensible interpretations of their results in the context of the original realistic situation.
- Make intelligent comments on the modelling assumptions and possible refinements to the model.
- Comprehend or understand the meaning of almost all translations into mathematics of common realistic contexts.
- Correctly refer the results of calculations back to given context and usually make sensible comments or predictions.
- Can distil the essential mathematical information from extended pieces of prose having mathematical content.
- Comment meaningfully on the mathematical information.
- Make appropriate and efficient use of contemporary calculator technology and other permitted resources, and are aware of any limitations to their use.
- Present results to an appropriate degree of accuracy.

Grade C candidates

- Recall or recognise most of the mathematical facts, concepts and techniques that are needed, and usually select appropriate ones to use in a variety of contexts.
- Manipulate mathematical expressions and use graphs, sketches and diagrams, all with a reasonable level of accuracy and skill.
- Use mathematical language with some skill and sometimes proceed logically through extended arguments or proofs.
- When confronted with unstructured problems they sometimes devise and implement an effective and efficient solution strategy.
- Occasionally notice and correct errors in their calculations.

- Recall or recognise most of the standard models that are needed and usually select appropriate ones to represent a variety of situations in the real world.
- Often correctly refer results from calculations using the model to the original situation; they sometimes give sensible interpretations of their results in context of the original realistic situation.
- Sometimes make intelligent comments on the modelling assumptions and possible refinements to the model.
- Comprehend or understand the meaning of most translations into mathematics of common realistic contexts.
- Often correctly refer the results of calculations back to the given context and sometimes make sensible comments or predictions.
- Distil much of the essential mathematical information from extended pieces of prose having mathematical content.
- Give some useful comments on this mathematical information.
- Usually make appropriate and effective use of contemporary calculator technology and other permitted resources, and are sometimes aware of any limitations to their use.
- Usually present results to an appropriate degree of accuracy.

Grade E candidates

- Recall or recognise some of the mathematical facts, concepts and techniques that are needed, and sometimes select appropriate ones to use in some contexts.
- Manipulate mathematical expressions and use graphs, sketches and diagrams, all with some accuracy and skill.
- Sometimes use mathematical language correctly and occasionally proceed logically through extended arguments or proofs.
- Recall or recognise some of the standard models that are needed and sometimes select appropriate ones to represent a variety of situations in the real world.
- Sometimes correctly refer results from calculations using the model to the original situation; they try to interpret their results in the context of the original realistic situation.
- Sometimes comprehend or understand the meaning of translations into mathematics of common realistic contexts.
- Sometimes correctly refer the results of calculations back to the given context and attempt to give comments or predictions.
- Distil some of the essential mathematical information from extended pieces of prose having mathematical content; they attempt to comment on this mathematical information.
- Candidates often make appropriate and efficient use of contemporary calculator technology and other permitted resources.
- Often present results to an appropriate degree of accuracy.

The table below shows how your uniform standardised mark is translated.

uniform mark (max. 100)	80	70	60	50	40
grade	A	B	C	D	E

The A* will be awarded to students who have achieved a grade A overall (480 UMS or more) and 180 UMS or more on the total of their Core 3 and Core 4 units. It is awarded for A Level qualification only and not for the AS qualification or individual units.

Four steps to successful revision

Step 1: Understand

- Study the topic to be learned slowly. Make sure you understand the logic or important concepts.
- Mark up the text if necessary – underline, highlight and make notes.
- Re-read each paragraph slowly.

GO TO STEP 2

Step 2: Summarise

- Now make your own revision note summary:
  What is the main idea, theme or concept to be learned?
  What are the main points? How does the logic develop?
  Ask questions: Why? How? What next?
- Use bullet points, mind maps, patterned notes.
- Link ideas with mnemonics, mind maps, crazy stories.
- Note the title and date of the revision notes
  (e.g. Mathematics: Differentiation, 3rd March).
- Organise your notes carefully and keep them in a file.

This is now in **short-term memory**. You will forget 80% of it if you do not go to Step 3.
GO TO STEP 3, but first take a 10 minute break.

Step 3: Memorise

- Take 25 minute learning 'bites' with 5 minute breaks.
- After each 5 minute break test yourself:
  Cover the original revision note summary
  Write down the main points
  Speak out loud (record yourself)
  Tell someone else
  Repeat many times.

The material is well on its way to **long-term memory**.
You will forget 40% if you do not do step 4. **GO TO STEP 4**

Step 4: Track/Review

- Create a Revision Diary (one A4 page per day).
- Make a revision plan for the topic, e.g. 1 day later, 1 week later, 1 month later.
- Record your revision in your Revision Diary, e.g.
  Mathematics: Differentiation, 3rd March 25 minutes
  Mathematics: Differentiation, 5th March 15 minutes
  Mathematics: Differentiation, 3rd April 15 minutes
  ... and then at monthly intervals.

Chapter 1

Core 3 and Core 4 (Pure Mathematics)

The following topics are covered in this chapter:

- Algebra and series
- Functions including exponential and logarithmic
- Coordinate geometry in the (x, y) plane
- Trigonometry

- Differentiation
- Integration
- Numerical methods
- Vectors

1.1 Algebra and series

LEARNING SUMMARY

After studying this section you should be able to:

- simplify rational expressions
- express rational expressions in partial fractions
- use the binomial expansion $(1 + x)^n$ for any rational n

Rational expressions

Core 3

A **rational expression** is of the form $\dfrac{f(x)}{g(x)}$, where $f(x)$ and $g(x)$ are polynomials in x.

To simplify a rational expression, factorise both $f(x)$ and $g(x)$ as far as possible, then cancel any factors that appear in both the numerator and denominator.

Examples

(a) $\dfrac{2x^2 + 6x}{2x^2 + 7x + 3} = \dfrac{2x(x + 3)}{(2x + 1)(x + 3)}$

$= \dfrac{2x}{2x + 1}$ This cannot be cancelled any further.

Factorise $x^2 - 4$ using difference between two squares.

(b) $\dfrac{3x^2 - 12}{2 - x} = \dfrac{3(x^2 - 4)}{2 - x}$

$x - 2 = -(2 - x)$.

$= \dfrac{3(x - 2)(x + 2)}{(2 - x)}$

$= -3(x + 2)$

When adding and subtracting algebra fractions, always use the lowest common denominator, for **example**,

First factorise the denominators.

The lowest common multiple of the denominators is $(x + 3)(x-4)(x - 3)$.

$\dfrac{x - 2}{x^2 - x - 12} + \dfrac{4}{x^2 - 9} = \dfrac{x - 2}{(x + 3)(x - 4)} + \dfrac{4}{(x + 3)(x - 3)}$

$= \dfrac{(x - 3)(x - 2) + 4(x - 4)}{(x + 3)(x - 4)(x - 3)}$

$= \dfrac{x^2 - 5x + 6 + 4x - 16}{(x + 3)(x - 4)(x - 3)}$

$= \dfrac{x^2 - x - 10}{(x + 3)(x - 4)(x - 3)}$

Key points from AS

Remainder theorem page 37
Algebraic division page 37

Partial fractions

Core 4

It is sometimes possible to decompose a rational function into **partial fractions**. This format is often useful when differentiating, integrating or expanding functions as a series. Look out for the following types of partial fractions.

Denominator contains linear factors only

$$\frac{x+7}{(x-2)(x+1)} \equiv \frac{A}{x-2} + \frac{B}{x+1} = \frac{A(x+1)+B(x-2)}{(x-2)(x+1)}$$

The symbol ≡ indicates that this is an identity. It is true for all values of x.

$$\Rightarrow x+7 \equiv A(x+1) + B(x-2)$$

These substitutions are chosen so that each factor in turn becomes zero.

Let $x = -1$, then $6 = -3B \Rightarrow B = -2$
Let $x = 2$, then $9 = 3A \Rightarrow A = 3$

Examination questions could have up to three linear factors in the denominator.

$$\therefore \frac{x+7}{(x-2)(x+1)} \equiv \frac{3}{x-2} - \frac{2}{x+1}$$

Denominator contains repeated linear factors

Do not use $(x+3)(x+1)(x+1)^2$ as the common denominator as it is not the lowest common multiple.

$$\frac{6x^2+15x+7}{(x+3)(x+1)^2} \equiv \frac{A}{x+3} + \frac{B}{x+1} + \frac{C}{(x+1)^2}$$

$$\equiv \frac{A(x+1)^2 + B(x+3)(x+1) + C(x+3)}{(x+3)(x+1)^2}$$

$$\Rightarrow 6x^2+15x+7 \equiv A(x+1)^2 + B(x+3)(x+1) + C(x+3)$$

Use a mixture of substitution and equating coefficients to find A, B and C.

Let $x = -1$, then $-2 = 2C \Rightarrow C = -1$
Let $x = -3$, then $16 = 4A \Rightarrow A = 4$
Equate x^2 terms: $6 = A + B$ and since $A = 4$, $B = 2$.

$$\therefore \frac{6x^2+15x+7}{(x+3)(x+1)^2} \equiv \frac{4}{x+3} + \frac{2}{x+1} - \frac{1}{(x+1)^2}$$

The degree of a polynomial is its highest power of x, for example $x^3 + 2x^2 - 3x + 4$ has degree 3.

Only **proper** fractions (when the numerator is of lower degree than the denominator) can be written in partial fraction form. For **improper fractions**, divide the denominator into the numerator first. This can be done by long division or by using identities.

Binomial expansion for rational n

Core 4

When n is rational, $(1+x)^n$ can be written as a series, using the binomial expansion

$$(1+x)^n = 1 + nx + \frac{n(n-1)}{2!}x^2 + \frac{n(n-1)(n-2)}{3!}x^3 + \dots$$

KEY POINT

Key points from AS

- **Binomial expansion when n is a positive integer**
  Revise AS page 42

If n is a positive integer, the series terminates at the term in x^n.

For all other values of n, the series is infinite and converges provided that $|x| < 1$.

Example

(a) Expand $\sqrt{1-4x}$ as an ascending series in x, as far as the term in x^3, giving the set of values of x for which the series is valid.

(b) By substituting $x = 0.01$ into the series expansion of $\sqrt{1-4x}$, find $\sqrt{96}$ correct to 4 decimal places.

> $1 - 4x = 1 + (-4x)$, so use $(1 + x)^n$ with $n = \frac{1}{2}$ and substituting $(-4x)$ for x.

> You should also state the set of values of x for which the expansion is valid, even when it is not specifically requested in the question.

(a) $\sqrt{1-4x} = (1-4x)^{\frac{1}{2}}$

$$= 1 + \tfrac{1}{2}(-4x) + \frac{(\frac{1}{2})(-\frac{1}{2})}{2!}(-4x)^2 + \frac{(\frac{1}{2})(-\frac{1}{2})(-\frac{3}{2})}{3!}(-4x)^3 + \dots$$

$$= 1 - 2x - 2x^2 - 4x^3 + \dots$$

The series is valid provided that $|4x| < 1$, i.e. $|x| < \frac{1}{4}$.

(b) Substituting $x = 0.01$ into $\sqrt{1-4x}$ gives $\sqrt{0.96} = \sqrt{\dfrac{96}{100}} = \dfrac{\sqrt{96}}{10}$.

So $\sqrt{96} = 10(1 - 2(0.01) - 2(0.01)^2 - 4(0.01)^3 + \dots)$

$$= 10(1 - 0.02 - 0.0002 - 0.000004 +)$$

$$= 9.7980 \text{ (4 d.p.)}$$

The function to be expanded must be in the form $(1 + \dots)^n$ or $(1 - \dots)^n$.
You may have to do some careful algebraic manipulation to get it in this form.

Example

Give the first four terms in the series expansion of $\dfrac{1}{(2+x)}$.

$$\frac{1}{(2+x)} = \frac{1}{2\left(1 + \dfrac{x}{2}\right)}$$

$$= \frac{1}{2}\left(1 + \frac{x}{2}\right)^{-1}$$

> Alternatively,
> $(2+x)^{-1} = 2^{-1}\left(1 + \dfrac{x}{2}\right)^{-1}$.

$$= \frac{1}{2}\left(1 + (-1)\left(\frac{x}{2}\right) + \frac{(-1)(-2)}{2!}\left(\frac{x}{2}\right)^2 + \frac{(-1)(-2)(-3)}{3!}\left(\frac{x}{2}\right)^3 + \dots\right)$$

$$= \frac{1}{2}\left(1 - \frac{x}{2} + \frac{x^2}{4} - \frac{x^3}{8} + \dots\right)$$

$$= \frac{1}{2} - \frac{x}{4} + \frac{x^2}{8} - \frac{x^3}{16} + \dots$$

Restriction on x: $\left|\dfrac{x}{2}\right| < 1$, i.e. $|x| < 2$.

Progress check

Core 3

1 Simplify:

(a) $\dfrac{2x^2 + 3x}{2x^2 + x - 3}$ (b) $\dfrac{9x - x^3}{3x^2 + x^3}$

Core 3

2 Express as a single fraction in its simplest form:

(a) $\dfrac{3(x - 4)}{(x + 2)(x - 1)} + \dfrac{2(x + 1)}{x - 1}$

(b) $\dfrac{x^2 - 3x + 2}{x^2 + 3x - 4} - \dfrac{1}{(x + 4)^2}$

Core 3

3 $\dfrac{4x^2 + x + 1}{(x + 2)(x - 3)(x + 1)} \equiv \dfrac{A}{x + 2} + \dfrac{B}{x - 3} + \dfrac{C}{x + 1}$

Find the values of A, B and C.

Core 4

4 Express in partial fractions:

$\dfrac{3x^2 + 4x + 5}{(x - 1)^2(x + 2)}$

Core 4

5 Write as a series in ascending powers of x as far as the term in x^3, stating the values of x for which the expansion is valid:

(a) $\dfrac{1}{1 + x}$ (b) $(1 - 3x)^{-2}$ (c) $\sqrt{4 + x}$

5 (c) $2 + \frac{1}{4}x - \frac{1}{64}x^2 + \frac{1}{512}x^3 + \cdots$ for $|x| < 4$

(b) $1 + 6x + 27x^2 + 108x^3 + \cdots$ for $|x| < \frac{1}{3}$

(a) $1 - x + x^2 - x^3 + \cdots$ for $|x| < 1$

4 $\dfrac{2}{x - 1} + \dfrac{4}{(x - 1)^2} + \dfrac{1}{x + 2}$

3 $A = 3$, $B = 2$, $C = -1$

2 (a) $\dfrac{2x^2 + 9x - 8}{(x + 2)(x - 1)}$ (b) $\dfrac{x^2 + 2x - 9}{(x + 4)^2}$

1 (a) $\dfrac{x - 1}{x}$ (b) $\dfrac{x}{3 - x}$

17

1.2 Functions including exponential and logarithmic

After studying this section you should be able to:

- *use function notation and understand domain and range*
- *understand composite functions and find the inverse of a one-one function*
- *sketch the graph of the inverse of a function from the graph of the function*
- *understand the properties of the modulus function*
- *use transformations to sketch graphs of related functions*
- *understand the definitions of e^x and $\ln x$*

LEARNING SUMMARY

Functions

Core 3

A **function** may be thought of as a rule which takes each member x of a set and assigns, or **maps**, it to some value y known as its **image**.

> x maps to y
> y is the image of x.

$$x \longrightarrow \boxed{\text{Function}} \longrightarrow y$$

> $f(x)$ is read as 'f of x'.

A letter such as f, g or h is often used to stand for a function. The function which squares a number and adds on 5, for example, can be written as $f(x) = x^2 + 5$. The same notation may also be used to show how a function affects particular values. For this function, $f(4) = 4^2 + 5 = 21$, $f(-10) = (-10)^2 + 5 = 105$ and so on.

An alternative notation for the same function is $f: x \mapsto x^2 + 5$.

> **KEY POINT**
>
> The set of values on which the function acts is called the **domain** and the corresponding set of image values is called the **range**.

The function $f(x) = x + 2$ with domain {3, 6, 10} is shown below.

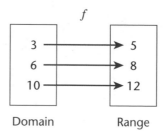

Domain   Range

> The symbol $\mathbb{R}$ is often used to stand for the complete set of numbers on the number line. This is the set of Real numbers.

The domain of a function may be an infinite set such as $\mathbb{R}$ but, in some cases, particular values must be omitted for the function to be valid.

For example, the function $f(x) = \dfrac{x + 2}{x - 3}$ cannot have 3 in its domain since division by zero is undefined.

Under a function, every member of the domain must have only one image. But, it is possible for different values in the domain to have the same image.

> There are different types of function.

> **KEY POINT**
>
> If for each element y in the range, there is *a unique* value of x such that $f(x) = y$ then f is a **one–one** function. If for any element y in the range, there is more than one value of x satisfying $f(x) = y$ then f is **many–one**.

Examples

The function $f(x) = 2x$ with domain $\mathbb{R}$ is one–one.

The function $f(x) = x^2$ with domain $\mathbb{R}$ is many–one.

The function $f(x) = x^2$ with domain $x > 0$ is one–one.

The function $f(x) = \sin x$ with domain $0° \leqslant x \leqslant 360°$ is many–one.

The function $f(x) = \sin x$ with domain $-90° \leqslant x \leqslant 90°$ is one–one.

> These examples show the importance of including the domain of the function in its definition.

Composition of functions

Core 3

The diagram shows how two functions f and g may be combined.

$$x \xrightarrow{\quad f \quad} f(x) \xrightarrow{\quad g \quad} g(f(x))$$

In this case, f is applied first to some value x giving $f(x)$. Then g is applied to the value $f(x)$ to give $g(f(x))$. This is usually written as $gf(x)$ and gf can be thought of as a new **composite function** defined from the functions f and g.

For example, if $f(x) = 3x$, $x \in \mathbb{R}$ and $g(x) = x + 2$, $x \in \mathbb{R}$ then
$$gf(x) = g(3x) = 3x + 2.$$

The order in which the functions are applied is important. The composite function fg is found by applying g first and then f.

In this case, $fg(x) = f(x + 2) = 3x + 6$.

Generally speaking, when two functions f and g are defined, the composite functions fg and gf will not be the same.

The inverse of a function

Core 3

The **inverse of a function** f is a function, usually written as f^{-1}, that *undoes* the effect of f. So the inverse of a function which adds 2 to every value, for example, will be a function that subtracts 2 from every value.

This can be written as $f(x) = x + 2$, $x \in \mathbb{R}$ and $f^{-1}(x) = x - 2$, $x \in \mathbb{R}$.

The domain of f^{-1} is given by the range of f.

Notice that $f^{-1}f(x) = f^{-1}(x + 2) = x$ and that $ff^{-1}(x) = f(x - 2) = x$.

A function can be either one–one or many–one, but only functions that are one–one can have an inverse. The reason is, that reversing a many–one function would give a mapping that is one–many, and this cannot be a function.

The diagram shows a many–one function. It does not have an inverse.

> Every value in the domain of a function can have only one image. This is an important property of functions.

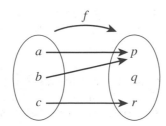

> If you reverse this diagram then p would have more than one image.

> Some important examples of restricting the domain of a function so that it will have an inverse are found in trigonometry. See page 28.

You can turn a many–one function into a one–one function by restricting its domain.

For example, the function $f(x) = x^2$, $x \in \mathbb{R}$ is many–one and so it does not have an inverse. However, the function $f(x) = x^2$, $x > 0$ is one–one and $f^{-1}(x) = +\sqrt{x}$, $x > 0$.

Fnding the inverse of a one–one function

Core 3

One way to find the inverse of a one–one function f is to write $y = f(x)$ and then rearrange this to make x the subject so that $x = f^{-1}(y)$. The inverse function is then usually defined in terms of x to give $f^{-1}(x)$. The domain of f^{-1} is the range of f.

> You can turn $f^{-1}(y)$ into $f^{-1}(x)$ by replacing every y with an x.

$$x \underset{f^{-1}}{\overset{f}{\rightleftarrows}} y$$

Example Find the inverse of the function $f(x) = \dfrac{x+2}{x-3}$, $x \neq 3$.

> Start by writing $y = f(x)$.

Define $\quad y = \dfrac{x+2}{x-3}$

> Rearrange to find x.

then $\quad y(x-3) = x+2$

> Multiply out the brackets.

$$xy - 3y = x + 2$$

> Collect the x terms on one side of the equation.

$$xy - x = 3y + 2$$

$$x(y-1) = 3y + 2$$

$$x = \frac{3y+2}{y-1}.$$

> $x = f^{-1}(y)$ so this defines the inverse function in terms of y.

> It is usual to express the inverse function in terms of x.

So $\quad f^{-1}(x) = \dfrac{3x+2}{x-1}$, $\quad x \neq 1$.

> The denominator cannot be allowed to be zero so 1 is not in the domain of f^{-1}.

Here are the graphs of $y = \dfrac{x+2}{x-3}$ and its inverse $y = \dfrac{3x+2}{x-1}$

> The graph of the inverse function is the reflection of the original graph in the line $y = x$. This is true for any inverse function but the same scale must be used on both axes or the effect is distorted.

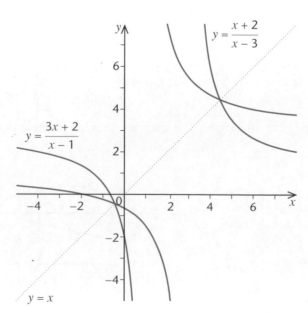

The modulus function

Core 3

The notation $|x|$ is used to stand for the modulus of x. This is defined as

$$|x| = \begin{cases} x & \text{when } x \geqslant 0 \quad \text{(when } x \text{ is positive, } |x| \text{ is just the same as } x\text{).} \\ -x & \text{when } x < 0 \quad \text{(when } x \text{ is negative, } |x| \text{ is the same as } -x\text{).} \end{cases}$$

The expression $|x-a|$ can be interpreted as the distance between the numbers x and a on the number line. In this way, the statement $|x-a| < b$ means that the distance between x and a is less than b.

It follows that $a - b < x < a + b$.

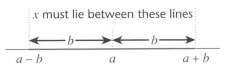

The graph of $y = |x|$

It follows that the graph of $y = |x|$ is the same as the graph of $y = x$ for positive values of x. But, when x is negative, the corresponding part of the graph of $y = x$ must be reflected in the x-axis to give the graph of $y = |x|$.

> $|x|$ is never negative so the graph of $y = |x|$ doesn't go below the x-axis anywhere.

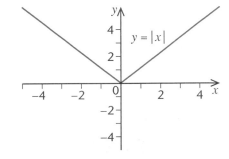

The graph of $y = |f(x)|$

The graph of $y = |f(x)|$ is the same as the graph of $y = f(x)$ for positive values of $f(x)$. But, when $f(x)$ is negative, the corresponding part of the graph of $y = f(x)$ must be reflected in the x-axis to give the graph of $y = |f(x)|$.

> The graph of $y = 1/x$ goes below the x-axis for negative values of x.

The diagram shows the graph of $y = \left|\dfrac{1}{x}\right|$.

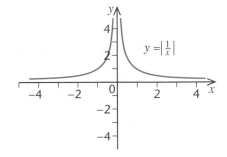

Transforming graphs

Core 3

The graph of some new function can often be obtained from the graph of a known function by applying a transformation. A summary of the standard transformations is given in the table.

> You may need to apply a combination of transformations in some cases.

Known function	New function	Transformation
$y = f(x)$	$y = f(x) + a$	Translation through a units parallel to y-axis.
	$y = f(x - a)$	Translation through a units parallel to x-axis.
	$y = af(x)$	One-way stretch with scale factor a parallel to the y-axis.
	$y = f(ax)$	One-way stretch with scale factor $\frac{1}{a}$ parallel to the x-axis.

Examples

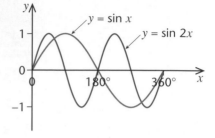

$y = af(x)$ does not move points on the x-axis.

Sometimes you may have to perform more than one transformation.

Example

Key points from AS

Circles
Revise AS page 39

Describe how to transform the circle $x^2 + y^2 = 25$, with centre (0, 0) and radius 5, to the circle with centre (3, −8) and radius 5. Write down the new equation.

Translate 3 units parallel to the x axis (to the right) and −8 units parallel to the y axis (down).

$$x^2 + y^2 = 25 \longrightarrow (x - 3)^2 + (y + 8)^2 = 25$$

The exponential function e^x

Core 3

An **exponential function** is one where the variable is a power or exponent. For example, any function of the form $f(x) = a^x$, where a is a constant, is an exponential function.

Exponential functions are often used to represent, or model, patterns of:
• growth when $a > 1$
• decay when $a < 1$.

Key points from AS

Exponential functions
Revise AS page 49

All scientific calculators will display values of e^x and 10^x. To display the value of e, for example, find e^1.

The diagram shows some graphs of exponential functions for different values of a.

All of the graphs pass through the point (0,1) and each one has a different gradient at this point. The value of the gradient depends on a.

There is a particular value of a for which the curve has gradient 1 at (0,1). The letter e is used to stand for this value and the function $f(x) = e^x$ is often called **the exponential function** because of its special importance to the subject.

e = 2.718281828 ... It is an irrational number and so it cannot be written exactly.

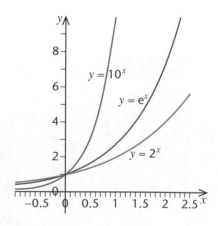

You may be asked to sketch other curves relating to e^x, for example

Notice $e^x > 0$ for all values of x.

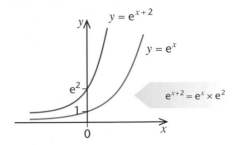

Reflection in y-axis.

Translation 2 units left or stretch, factor e^2, parallel to y-axis.

$e^{x+2} = e^x \times e^2$

The natural logarithmic function, In x

Core 3

Key points from AS

Logarithms Revise AS page 49

If $y = e^x$ then $x = \log_e y$. This is usually written as $x = \ln y$.

Logarithms to the base e are called *natural logarithms*.

Your calculator has a key for natural logs, probably labelled ln.

$e^0 = 1 \Rightarrow \ln 1 = 0$
$e^1 = e \Rightarrow \ln e = 1$

In fact, $\ln e^{f(x)} = f(x)$, for example $\ln e^{x^2} = x^2$, $\ln e^{\sin x} = \sin x$.

Since the exponential and natural logarithm functions are inverses of each other, their graphs are symmetrical about the line $y = x$.

This is true for logs of any base.

The diagram shows that the domain of the natural logarithm function is given by $x > 0$.

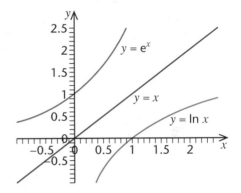

You may be asked to sketch other curves relating to $y = \ln x$, for example

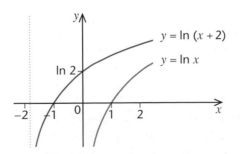

Translation 2 units to the left.

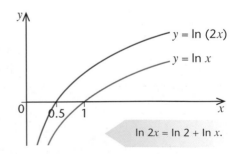

$\ln 2x = \ln 2 + \ln x$.

Stretch parallel to the x-axis factor $\frac{1}{2}$ or translation $\ln 2$ units up.

Progress check

Core 3

1 $f(x) = x^2 + 5$ and $g(x) = 2x + 3$.

    (a) Write $fg(x)$ in terms of x.   (b) Find $fg(10)$.

    (c) Find the values of x for which $fg(x) = gf(x)$.

Core 3

2 Find the inverse of the function $f(x) = \dfrac{x+5}{x-2}$, $x \in \mathbb{R}$, $x \neq 2$ and state the domain of the inverse function.

Core 3

3 (a) Sketch $y = (x-2)(x+2)$.

    (b) Sketch $y = |x^2 - 4|$.

Core 3

4 Solve $|x-2| < 5$.

Core 3

5 The diagram shows $y = f(x)$

On separate diagrams, sketch

    (a) $y = f(x) + 1$

    (b) $y = f(x+1)$

    (c) $y = f(2x)$

    (d) $y = 2f(x)$

    (e) $y = f(\tfrac{1}{2}x)$

Core 3

6 On the same diagram, sketch $y = x^2$, $y = (x-2)^2$, $y = -x^2$.

Core 3

7 On the same diagram, sketch $y = e^x$, $y = e^{2x}$, $y = -e^x$.

4 $-3 < x < 7$

2 $f(x) = \dfrac{2x+5}{x-1}$, $x \in \mathbb{R}$, $x \neq 1$.

1 (a) $fg(x) = 4x^2 + 12x + 14$ (b) $fg(10) = 534$ (c) $x = -0.08452, -5.915$.

1.3 Coordinate geometry in the (x, y) plane

After studying this section you should be able to:

- *sketch curves in Cartesian and parametric form*
- *convert the equation of a curve between Cartesian and parametric form*

LEARNING SUMMARY

Cartesian and parametric form

Core 4

Key points from AS

- **Curve sketching**
  Revise AS page 51

Cartesian form

A curve in Cartesian form is defined in terms of x and y only, **for example**
$$y = x^2 + 3x \text{ in which } y \text{ is given } \textbf{explicitly} \text{ in terms of } x.$$
$$x^2 + 2xy + y^2 = 9 \text{ in which the equation is given } \textbf{implicitly}.$$

To sketch a function given in Cartesian form:

If all the powers of x are even, the graph is symmetrical in the y-axis.

1 Find where the curve **crosses the axes** by putting $x = 0$ and $y = 0$.
2 Check for **symmetry**.
3 Check for **stationary points**.

Find when $\dfrac{dy}{dx} = 0$ and then investigate.

Look for values that make the denominator zero.

4 Check what happens for **large values of x and y**.
5 Check for **discontinuities**.

Parametric form

The parameter θ is often used when trigonometric functions are involved.

A curve is defined in **parametric form** by expressing x and y in terms of a third variable, for example $x = f(t)$, $y = g(t)$. In this case, t is the parameter.

The curve is plotted by finding the values of x and y for various values of the parameter.

Example

Plot the curve $x = 2t^2$, $y = 4t$ for $-3 \leqslant t \leqslant 3$ and find the Cartesian equation of the curve.

The curve is a parabola.

t	−3	−2	−1	0	1	2	3
x	18	8	2	0	2	8	18
y	−12	−8	−4	0	4	8	12

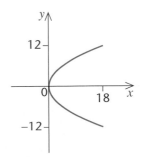

Eliminate t to obtain the Cartesian equation.

$$y = 4t \Longrightarrow t = \frac{y}{4}.$$

Substituting into $x = 2t^2$ gives $x = 2\left(\dfrac{y}{4}\right)^2 = \dfrac{y^2}{8}$ so $y^2 = 8x$.

Key points from AS

- **Trigonometric identities**
  Revise AS page 47

When the parameter is θ, it may be necessary to use a trigonometric identity to find the Cartesian equation.

Key points from AS

- **Circle**
  Revise AS page 39

For example, when $x = a \cos \theta$ and $y = a \sin \theta$, use $\cos^2 \theta + \sin^2 \theta = 1$. This gives
$\dfrac{x^2}{a^2} + \dfrac{y^2}{a^2} = 1$ or $x^2 + y^2 = a^2$.

This is a circle, centre (0, 0), radius a.

Also, when $x = a \cos \theta$, $y = b \sin \theta$, $\dfrac{x^2}{a^2} + \dfrac{y^2}{b^2} = 1$.

This curve is an ellipse.

Progress check

Core 4

1 A curve is defined parametrically by $x = 2t$, $y = \dfrac{2}{t}$.

   (a) Give the Cartesian equation of the curve.

   (b) Sketch the curve.

Core 4

2 Describe the curve defined parametrically by $x = 5 \cos \theta$, $y = 5 \sin \theta$ and give its Cartesian equation.

2 Circle, centre (0, 0), radius 5; $x^2 + y^2 = 25$.

(b)

1 (a) $xy = 4$ (hyperbola)

1.4 Trigonometry

Secant, cosecant and cotangent

Core 3

Key points from AS

- **Trigonometry** *pages 44–48*

These trigonometric functions, commonly known as **sec**, **cosec** and **cot**, are defined from the more familiar sin, cos and tan functions as follows:

$$\sec x = \frac{1}{\cos x} \qquad \operatorname{cosec} x = \frac{1}{\sin x} \qquad \cot x = \frac{1}{\tan x}$$

Each function is periodic and has either line or rotational symmetry. The domain of each one must be restricted to avoid division by zero.

You need to be able to work in degrees or radians.

Remember $180° = \pi$ radians.

$y = \sec x$

- The period of sec is $360°$ (2π radians) to match the period of cos.
- Notice that $\sec x$ is undefined whenever $\cos x = 0$.
- The graph is symmetrical about every vertical line passing through a vertex.
- It has rotational symmetry of order 2 about the points in the x-axis corresponding to $90° \pm 180°n$ ($\frac{\pi}{2} \pm \pi n$).

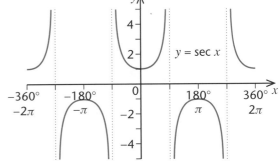

$y = \operatorname{cosec} x$

- The period of cosec is $360°$ (2π radians) to match the period of sin.
- Notice that $\operatorname{cosec} x$ is undefined whenever $\sin x = 0$.
- The graph is symmetrical about every vertical line passing through a vertex.
- It has rotational symmetry of order 2 about the points on the x-axis corresponding to $0°$, $\pm 180°$, $\pm 360°$, ..., $(0, \pm \pi, \pm 2\pi, ...)$.

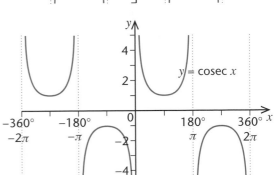

$y = \cot x$

$$\frac{1}{\tan x} = \frac{\cos x}{\sin x}.$$

- The period of cot is $180°$ to match the period of tan.
- Notice that $\cot x$ is undefined whenever $\sin x = 0$.
- The graph has rotational symmetry of order 2 about the points on the x-axis corresponding to $0°$, $\pm 90°$, $\pm 180°$, ..., $(0, \pm \frac{\pi}{2}, \pm \pi, ...)$.

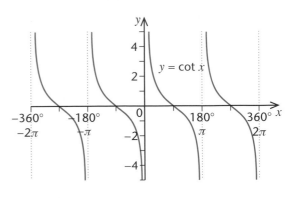

Inverse trigonometric functions

Core 3

See page 19 to find out about inverse functions.

The sine, cosine and tangent functions are all many–one and so do not have inverses on their full domains. However, it is possible to restrict their domains so that each one has an inverse. The graphs of these inverse functions are given below.

You need to know these graphs and to understand the need for restricting the domains.

$f(x) = \sin x, \ -\frac{\pi}{2} \leqslant x \leqslant \frac{\pi}{2}$
$(-90° \leqslant x \leqslant 90°)$

$\Rightarrow f^{-1}(x) = \sin^{-1} x, \ -1 \leqslant x \leqslant 1.$

$f(x) = \cos x, \ 0 \leqslant x \leqslant \pi$
$(0° \leqslant x \leqslant 180°)$

$\Rightarrow f^{-1}(x) = \cos^{-1} x, \ -1 \leqslant x \leqslant 1.$

$f(x) = \tan x, \ -\frac{\pi}{2} < x < \frac{\pi}{2}$
$(-90° < x < 90°)$

$\Rightarrow f^{-1}(x) = \tan^{-1} x, \ x \in \mathbb{R}.$

This is the notation used for inverse trig functions on your calculator.

But some texts use:

arcsin for $\sin^{-1}$
arccos for $\cos^{-1}$
arctan for $\tan^{-1}$

$\sin^{-1} x$ means:
'the angle whose sine is x'
e.g. $\sin^{-1} 0.5 = 30°$.

In radians $\sin^{-1} 0.5 = \frac{\pi}{6}$.

$\cos^{-1} x$ means:
'the angle whose cosine is x'

e.g. $\cos^{-1} 0.5 = 60°$.

In radians $\cos^{-1} 0.5 = \frac{\pi}{3}$.

$\tan^{-1} x$ means:
'the angle whose tangent is x'

e.g. $\tan^{-1} 1 = 45°$.

In radians $\tan^{-1} 1 = \frac{\pi}{4}$.

When you use the functions $\sin^{-1}$, $\cos^{-1}$ and $\tan^{-1}$ on your calculator, the value given is called the **principal value** (PV).

Trigonometric identities

Core 3

You should *learn* the identities listed below. It is so much easier to simplify expressions, establish new identities and solve equations if you have a firm grasp of these basic results.

The second result can be obtained from the first by dividing throughout by $\cos^2\theta$.
The third result can be obtained in a similar way, by dividing throughout by $\sin^2\theta$.

• The **Pythagorean identities** are:

$$\cos^2\theta + \sin^2\theta \equiv 1$$

$$1 + \tan^2\theta \equiv \sec^2\theta$$

$$\cot^2\theta + 1 \equiv \operatorname{cosec}^2\theta$$

• The **compound angle identities** for $(A + B)$ are:

$$\sin(A + B) \equiv \sin A \cos B + \cos A \sin B$$

$$\cos(A + B) \equiv \cos A \cos B - \sin A \sin B$$

$$\tan(A + B) \equiv \frac{\tan A + \tan B}{1 - \tan A \tan B}.$$

The corresponding results for $(A - B)$ are:

These results can be worked out from the ones for $(A + B)$. Just replace every '+' with a '−' and vice versa. This makes it easier to learn the results.

$$\sin(A - B) \equiv \sin A \cos B - \cos A \sin B$$

$$\cos(A - B) \equiv \cos A \cos B + \sin A \sin B$$

$$\tan(A - B) \equiv \frac{\tan A - \tan B}{1 + \tan A \tan B}.$$

You can obtain these results from the compound angle formulae by setting $A = B$.

- The **double angle identities** are:

$$\sin 2A \equiv 2 \sin A \cos A$$
$$\cos 2A \equiv \cos^2 A - \sin^2 A$$
$$\equiv 2 \cos^2 A - 1$$
$$\equiv 1 - 2 \sin^2 A$$
$$\tan 2A \equiv \frac{2 \tan A}{1 - \tan^2 A}.$$

These variations are very useful – it is worth knowing them.

- The **half angle identities** are:

$$\sin^2 \tfrac{1}{2} A \equiv \tfrac{1}{2}(1 - \cos A)$$
$$\cos^2 \tfrac{1}{2} A \equiv \tfrac{1}{2}(1 + \cos A).$$

These results will be useful for integration later in the course.

Proving identities

Core 3

The basic technique for proving a given identity is:

- Start with the expression on one side of the identity.
- Use known identities to replace some part of it with an equivalent form.
- Simplify the result and compare with the expression on the other side.

This is a skill. The more you do, the better you will become.

With practice you get to know which parts to replace so that the required result is established.

Example

Prove that $\dfrac{\cos 2A}{\cos A - \sin A} \equiv \cos A + \sin A.$

Starting with the LHS:

$$\frac{\cos 2A}{\cos A - \sin A} \equiv \frac{\cos^2 A - \sin^2 A}{\cos A - \sin A}$$

Using one of the double angle results.

$$\equiv \frac{(\cos A + \sin A)(\cos A - \sin A)}{\cos A - \sin A}$$

Using the difference of two squares.

$$\equiv \cos A + \sin A \equiv \text{RHS}.$$

Cancelling the common factor.

It is often a good idea to replace sec, cosec and cot with sin, cos and tan at the start.

Example

Prove that $\sec^2 A \equiv \dfrac{\operatorname{cosec} A}{\operatorname{cosec} A - \sin A}.$

It's often easier to start with the more complicated looking side and try to simplify it.

$$\text{RHS} \equiv \frac{\dfrac{1}{\sin A}}{\dfrac{1}{\sin A} - \sin A} \equiv \frac{1}{1 - \sin^2 A}$$

Replacing cosec A with $\dfrac{1}{\sin A}$ and multiplying the numerator and denominator by sin A.

$$\equiv \frac{1}{\cos^2 A} \equiv \sec^2 A \equiv \text{LHS}.$$

Using one of the Pythagorean identities and the definition of sec A.

Trigonometric equations

Core 3

Identities may be used to re-write an equation in a form that is easier to solve.

Example   Solve the equation $\cos 2x + \sin x + 1 = 0$ for $-\pi \leqslant x \leqslant \pi$.

Replacing $\cos 2x$ with $1 - 2\sin^2 x$ gives $1 - 2\sin^2 x + \sin x + 1 = 0$

$$\Rightarrow 2\sin^2 x - \sin x - 2 = 0.$$

> You need to recognise that this is a quadratic in $\sin x$.

Using the quadratic formula gives $\sin x = \dfrac{1 \pm \sqrt{1 + 16}}{4} = \dfrac{1 \pm \sqrt{17}}{4}$

$$\Rightarrow \sin x = 1.28 \ldots \text{ (no solutions) or } \sin x = -0.78077 \ldots.$$

Using $\sin^{-1}$ gives $\qquad x = -0.8959$ radians to 4 d.p.

> Set calculator to radian mode.

> $-1 \leqslant \sin x \leqslant 1$ for all values of x.

Another solution in the interval $-\pi \leqslant x \leqslant \pi$ is $-\pi + 0.8959$ giving $x = -2.2457$ radians to 4 d.p.

The diagram shows there are no other solutions in this interval.

An important method is that the expression $a \cos \theta + b \sin \theta$ can be written in any of the equivalent forms $r \cos(\theta \pm \alpha)$ or $r \sin(\theta \pm \alpha)$ for a suitable choice of r and α. The convention is to take $r > 0$.

KEY POINT

One application of the method is to solve equations of the form $a \cos \theta + b \sin \theta = c$.

Example   Express $3 \cos \theta - 4 \sin \theta$ in the form $r \cos(\theta + \alpha)$, $r > 0$. Use the result to solve the equation $3 \cos \theta - 4 \sin \theta = 2$ for $-\pi \leqslant \theta \leqslant \pi$.

> You need to expand $r \cos(\theta + \alpha)$ and match corresponding parts of the identity.

$$r \cos(\theta + \alpha) \equiv 3 \cos \theta - 4 \sin \theta$$
$$\Rightarrow r \cos \theta \cos \alpha - r \sin \theta \sin \alpha \equiv 3 \cos \theta - 4 \sin \theta$$
$$\Rightarrow r \cos \alpha = 3 \text{ and } r \sin \alpha = 4$$

> $\dfrac{r \sin \alpha}{r \cos \alpha} = \dfrac{4}{3}$
>
> $\Rightarrow \tan \alpha = \dfrac{4}{3}$

$$\Rightarrow r = \sqrt{3^2 + 4^2} = 5 \text{ and } \alpha = \tan^{-1}(\tfrac{4}{3}) = 0.9273^c \text{ to 4 d.p.}$$
$$\Rightarrow 3 \cos \theta - 4 \sin \theta \equiv 5 \cos(\theta + 0.9273^c).$$

Using this result in the equation $3 \cos \theta - 4 \sin \theta = 2$ gives

$$5 \cos(\theta + 0.9273^c) = 2$$
$$\Rightarrow \cos(\theta + 0.9273^c) = 0.4$$

This gives $\theta + 0.9273^c = \pm 1.1593^c \Rightarrow \theta = 0.232^c$ or $\theta = -2.087^c$ to 3 d.p.

Progress check

Core 3

1   Prove the identity $\sin 2\theta \equiv \dfrac{2 \tan \theta}{1 + \tan^2 \theta}$.

Core 3

2   Solve the equation $\cos 2x - \sin x = 0$ for $0° \leqslant x \leqslant 360°$.

Core 3

3   (a)   Write $5 \sin x + 12 \cos x$ in the form $r \sin(x + \alpha)$ where $0 < \alpha < \dfrac{\pi}{2}$.

    (b)   Solve the equation $5 \sin x + 12 \cos x = 7$ for $-\pi \leqslant x \leqslant \pi$.

3 (a) $13 \sin(x + 1.176^c)$ (b) $x = -0.607^c$ or $x = 1.397^c$ to 3 d.p.

2 $x = 30°, 150°$ or $270°$

1 RHS $= \dfrac{2 \tan \theta}{\sec^2 \theta} = 2 \tan \theta \times \cos^2 \theta = 2 \sin \theta \cos \theta = \sin 2\theta = $ LHS.

1.5 Differentiation

After studying this section you should be able to:

- use the chain rule to differentiate composite functions
- differentiate functions based on e^x and $\ln x$
- differentiate trigonometric functions
- differentiate using the product and quotient rules
- understand and use the relation $\dfrac{dy}{dx} = 1 \div \dfrac{dx}{dy}$
- find derivatives of functions defined parametrically and implicitly

LEARNING SUMMARY

The chain rule

Core 3

> **KEY POINT**
>
> The **chain rule** can be written as $\dfrac{dy}{dx} = \dfrac{dy}{du} \times \dfrac{du}{dx}$, where u is a function of x.

It shows how to differentiate a composite function by separating the process into simpler stages and combining the results.

Example   Differentiate   (a) $y = (3x + 2)^{10}$

(b) $y = \sqrt{x^2 - 9}$.

Key points from AS

- **Differentiation**
  Revise AS pages 28–30

(a) Taking $u = 3x + 2$ gives $y = u^{10}$ and $\dfrac{dy}{du} = 10u^9$, so $\dfrac{dy}{du} = 10(3x + 2)^9$.

It also gives $\dfrac{du}{dx} = 3$.

So, from the chain rule: $\dfrac{dy}{dx} = 10(3x + 2)^9 \times 3 = 30(3x + 2)^9$.

(b) Taking $u = x^2 - 9$ gives $y = \sqrt{u} = u^{\frac{1}{2}}$, and $\dfrac{dy}{du} = \frac{1}{2} u^{-\frac{1}{2}}$, so $\dfrac{dy}{du} = \frac{1}{2}(x^2 - 9)^{-\frac{1}{2}}$.

It also gives $\dfrac{du}{dx} = 2x$.

So, from the chain rule: $\dfrac{dy}{dx} = \frac{1}{2}(x^2 - 9)^{-\frac{1}{2}} \times 2x = \dfrac{x}{\sqrt{x^2 - 9}}$.

The chain rule can also be used to establish results for **connected rates of change**.

For example, the rate of change of the volume of a sphere can be written as $\dfrac{dv}{dt}$.

The corresponding rate of change of the radius of the sphere can be written as $\dfrac{dr}{dt}$. Using the chain rule, the connection between these rates of change is given by:

$$\frac{dv}{dt} = \frac{dv}{dr} \times \frac{dr}{dt}.$$

Also, 'since $v = \frac{4}{3} \pi r^3$, it follows that $\dfrac{dv}{dr} = 4\pi r^2$, so, the connection can now be written as $\dfrac{dv}{dt} = 4\pi r^2 \dfrac{dr}{dt}$.

The exponential function e^x

Core 3

The exponential function was defined on page 22. It has the special property that it is unchanged by differentiation. So, if $y = e^x$ then $\frac{dy}{dx} = e^x$.

Constant multiples of the exponential function behave in the same way, so if

$y = ae^x$ then $\frac{dy}{dx} = ae^x$ where a is any constant.

Examples
Differentiate with respect to x:

(a) $y = 2e^x$ (b) $y = x^3 + e^x$ (c) $f(x) = \frac{1}{x} - 3e^x$

(a) $\frac{dy}{dx} = 2e^x$ (b) $\frac{dy}{dx} = 3x^2 + e^x$ (c) $f'(x) = -\frac{1}{x^2} - 3e^x$

You may need to use the chain rule.

Examples
Differentiate with respect to x:

(a) $y = e^{3x+2}$ (b) $y = e^{4x^2}$

(a) Let $u = 3x + 2$ so that $y = e^u$

$\frac{du}{dx} = 3$ and $\frac{dy}{du} = e^u$

Using the chain rule

$\frac{dy}{dx} = \frac{dy}{du} \times \frac{du}{dx} = e^u \times 3 = 3\,e^{3x+2}$

(b) Let $u = 4x^2$, so that $y = e^u$

$\frac{du}{dx} = 8x$ and $\frac{dy}{du} = e^u$

$\frac{dy}{dx} = \frac{dy}{du} \times \frac{du}{dx} = e^u \times 8x = 8xe^{4x^2}$

In general, it can be shown using the chain rule

$f'(x)$ is the derivative, with respect to x, of $f(x)$.

If $y = e^{f(x)}$ then $\frac{dy}{dx} = f'(x)e^{f(x)}$.

KEY POINT

In particular, it is useful to remember that if $y = e^{ax+b}$, then $\frac{dy}{dx} = ae^{ax+b}$

Logarithmic functions $\ln x$ and $\ln f(x)$

Core 3

If $y = \ln x$ then $\frac{dy}{dx} = \frac{1}{x}$.

KEY POINT

Key points from AS

- **Logarithms**
  Revise AS page 49

$\log a + \log b = \log(ab)$.
$\log x^k = k \log x$.

This result is easily extended to functions of the form $y = \ln(kx)$ and $y = \ln(x^k)$ using the laws of logarithms.

If $y = \ln(kx)$, then $y = \ln k + \ln x$

giving $\dfrac{dy}{dx} = \dfrac{1}{x} + 0 = \dfrac{1}{x}$.

> Since $\ln k$ is a constant, its derivative is zero.

For example, if $y = \ln(5x)$, then $\dfrac{dy}{dx} = \dfrac{1}{x}$.

If $y = \ln(x^k)$, then $y = k \ln x$ giving $\dfrac{dy}{dx} = k \times \dfrac{1}{x} = \dfrac{k}{x}$.

For example, if $y = \ln\left(\dfrac{1}{x^2}\right) = \ln(x^{-2}) = -2 \ln x$, then $\dfrac{dy}{dx} = -2 \times \dfrac{1}{x} = -\dfrac{2}{x}$.

Always check whether a logarithmic function can be simplified before attempting to differentiate.

You may need to use the chain rule to differentiate $y = \ln f(x)$.

Example

> This function cannot be simplified.

Differentiate $y = \ln(x^2 + 3x)$.

Let $u = x^2 + 3x$ so that $y = \ln u$

$$\dfrac{du}{dx} = 2x + 3 \quad \text{and} \quad \dfrac{dy}{du} = \dfrac{1}{u} = \dfrac{1}{x^2 + 3x}$$

Using the chain rule

$$\dfrac{dy}{dx} = \dfrac{dy}{du} \times \dfrac{du}{dx} = \dfrac{1}{x^2 + 3x} \times (2x + 3)$$

> Notice that $2x + 3$ is the derivative of $x^2 + 3x$.

$$= \dfrac{2x + 3}{x^2 + 3x}$$

In general, it can be shown using the chain rule

> This can be used directly without showing the chain rule working.

$$\dfrac{d}{dx} \ln(f(x)) = \dfrac{f'(x)}{f(x)}.$$

KEY POINT

Progress check

Find $\dfrac{dy}{dx}$ in each of the following questions.

Core 3 1 $y = (4 - 3x)^6$

Core 3 2 $y = \sqrt{2x + 1}$

Core 3 3 $y = e^{3x+1} + 3x$

Core 3 4 $y = \ln(7x) + \ln(x^4)$

Core 3 5 $y = \ln(x^2 - 3)$

5 $\dfrac{2x}{x^2 - 3}$

4 $\dfrac{5}{x}$

3 $3e^{3x+1} + 3$

2 $(2x + 1)^{-\frac{1}{2}}$

1 $-18(4 - 3x)^5$

33

Trigonometric functions

Core 3

You should learn the derivatives of the three main **trigonometric functions**:

$$\frac{d}{dx}(\sin x) = \cos x \qquad \frac{d}{dx}\sin(ax+b) = a\cos(ax+b)$$

$$\frac{d}{dx}(\cos x) = -\sin x \qquad \frac{d}{dx}\cos(ax+b) = -a\sin(ax+b)$$

$$\frac{d}{dx}(\tan x) = \sec^2 x \qquad \frac{d}{dx}\tan(ax+b) = a\sec^2(ax+b)$$

> When differentiating, the angle must be in radians.

Example

Find the exact value of the gradient of $y = \cos 2x$ when $x = \dfrac{\pi}{6}$.

$$y = \cos 2x \implies \frac{dy}{dx} = -2\sin 2x$$

When $x = \dfrac{\pi}{6}$, $\dfrac{dy}{dx} = -2\sin\dfrac{\pi}{3} = -2 \times \dfrac{\sqrt{3}}{2} = -\sqrt{3}$

The **chain rule** can often be used to differentiate expressions involving trigonometric functions.

Example

Differentiate: (a) $y = 5\sin^4 x$ (b) $y = e^{\tan 4x}$ (c) $y = \ln(\sin x)$

> $\sin^4 x = (\sin x)^4$.

(a)   Let $u = \sin x$ so that $y = 5u^4$

Then $\dfrac{du}{dx} = \cos x$ and $\dfrac{dy}{du} = 20u^3 = 20(\sin x)^3 = 20\sin^3 x$

> With practice you should be able to do the working mentally.

By the chain rule:

$$\frac{dy}{dx} = \frac{dy}{du} \times \frac{du}{dx} = 20\sin^3 x \times (\cos x)$$

$$= 20\cos x \sin^3 x.$$

> Although not essential, it is usual to write an expression with the least complicated factors first.

(b)   Let $u = \tan 4x$ so that $y = e^u$

$$\frac{du}{dx} = 4\sec^2 4x \text{ and } \frac{dy}{du} = e^u = e^{\tan 4x}.$$

$$\therefore \quad \frac{dy}{dx} = e^{\tan 4x} \times 4\sec^2 4x$$

$$= 4\sec^2 4x\, e^{\tan 4x}.$$

> In general, by the chain rule:
> $$\frac{d}{dx}(e^{f(x)}) = f'(x)e^{f(x)}.$$

(c) Use the result $y = \ln f(x) \implies \dfrac{dy}{dx} = \dfrac{f'(x)}{f(x)}$

> $f(x) = \sin x$, $f'(x) = \cos x$.

$$y = \ln(\sin x)$$

$$\frac{dy}{dx} = \frac{\cos x}{\sin x}$$

$$= \cot x.$$

The product rule

Core 3

Expressions written as the product of two functions, $f(x) \times g(x)$, can be differentiated using the **product rule**.

> **KEY POINT**
>
> If $y = uv$, where $u = f(x)$ and $v = g(x)$, then $\dfrac{dy}{dx} = u \dfrac{dv}{dx} + v \dfrac{du}{dx}$.

Example

Differentiate with respect to x: (a) $y = x^2 \sin 4x$ (b) $y = x^3 e^{-x}$

With practice, the side working can be done mentally.

(a) Let $y = uv$, where $u = x^2$ and $v = \sin 4x$ Side working

$$\frac{dy}{dx} = u \frac{dv}{dx} + v \frac{du}{dx}$$

$$= x^2 \times 4 \cos 4x + \sin 4x \times 2x$$

$$= 2x(2x \cos 4x + \sin 4x)$$

$u = x^2 \implies \dfrac{du}{dx} = 2x$

$v = \sin 4x \implies \dfrac{dv}{dx} = 4 \cos 4x$

Tidy the answer by taking out factors if possible.

(b) Let $y = uv$, where $u = x^3$ and $v = e^{-x}$ Side working

$$\frac{dy}{dx} = x^3 \times (-e^{-x}) + e^{-x} \times 3x^2$$

$$= x^2 e^{-x}(-x + 3)$$

$u = x^3 \implies \dfrac{du}{dx} = 3x^2$

$v = e^{-x} \implies \dfrac{dv}{dx} = -e^{-x}$

The quotient rule

Core 3

Expressions written in the form $\dfrac{f(x)}{g(x)}$ can be differentiated using the **quotient rule**.

Before using the quotient rule, check whether the expression can be simplified first, such as in $y = \dfrac{4x^2 + 3x + 2}{x}$.

> **KEY POINT**
>
> If $y = \dfrac{u}{v}$ where $u = f(x)$ and $v = g(x)$, then $\dfrac{dy}{dx} = \dfrac{v \dfrac{du}{dx} - u \dfrac{dv}{dx}}{v^2}$.

Example

Find $\dfrac{dy}{dx}$ when $y = \dfrac{e^{2x}}{x^3}$.

Side working:

$$\frac{dy}{dx} = \frac{v \dfrac{du}{dx} - u \dfrac{dv}{dx}}{v^2}$$

$$= \frac{x^3(2e^{2x}) - e^{2x}(3x^2)}{(x^3)^2}$$

$$= \frac{x^2 e^{2x}(2x - 3)}{x^6}$$

$$= \frac{e^{2x}(2x - 3)}{x^4}$$

$u = e^{2x} \implies \dfrac{du}{dx} = 2e^{2x}$

$v = x^3 \implies \dfrac{dv}{dx} = 3x^2$

Take out factors.

Simplify.

Example

Find the coordinates of the stationary point on the curve $y = \dfrac{\ln x}{x}$.

Key points from AS

• **Stationary points**
  Revise AS page 51

Side working:

$$\frac{dy}{dx} = \frac{x \times \dfrac{1}{x} - \ln x \times 1}{x^2}$$

$$= \frac{1 - \ln x}{x^2}$$

$u = \ln x \Rightarrow \dfrac{du}{dx} = \dfrac{1}{x}$

$v = x \Rightarrow \dfrac{dv}{dx} = 1$

$\ln e = 1$.

$\dfrac{dy}{dx} = 0$ when $\ln x = 1$, i.e. when $x = e$

When $x = e$, $y = e^{-1}$, so the stationary point is at (e, e^{-1}).

Alternative formats of product and quotient rules when $u = f(x)$ and $v = g(x)$:

These formats are sometimes quoted. Learn and use the ones you find easiest to remember.

Product rule: If $y = f(x) \times g(x)$, then $\dfrac{dy}{dx} = f'(x)g(x) + f(x)g'(x)$

Quotient rule: If $y = \dfrac{f(x)}{g(x)}$, then $\dfrac{dy}{dx} = \dfrac{f'(x)g(x) - f(x)g'(x)}{[g(x)]^2}$

Using the result $\dfrac{dy}{dx} = 1 \div \dfrac{dx}{dy}$

Core 3

$$\frac{dy}{dx} = 1 \div \frac{dx}{dy} = \frac{1}{\dfrac{dx}{dy}}.$$

KEY POINT

For example, if $x = 4y + 2$, then $\dfrac{dx}{dy} = 4$ and $\dfrac{dy}{dx} = \dfrac{1}{\dfrac{dx}{dy}} = \dfrac{1}{4}$.

The relationship is very useful for differentiating inverse trigonometrical functions.

Example

This could be written:
Find $\dfrac{dy}{dx}$ when $x = \sin y$.

Find $\dfrac{dy}{dx}$ when (a) $y = \sin^{-1} x$ (b) $y = \tan^{-1} x$

(a) $y = \sin^{-1} x \Rightarrow x = \sin y$

Use $\cos^2 y + \sin^2 y = 1$.

$$\frac{dx}{dy} = \cos y = \sqrt{1 - \sin^2 y} = \sqrt{1 - x^2}$$

$$\frac{dy}{dx} = \frac{1}{\dfrac{dx}{dy}} = \frac{1}{\sqrt{1 - x^2}}$$

(b) $\quad y = \tan^{-1} x \Rightarrow x = \tan y$

$$\frac{dx}{dy} = \sec^2 y = 1 + \tan^2 y = 1 + x^2$$

$$\frac{dy}{dx} = \frac{1}{\dfrac{dx}{dy}} = \frac{1}{1 + x^2}$$

> Use $1 + \tan^2 y = \sec^2 y.$

> **Core 4**

It is also useful for differentiating $y = a^x$ as follows.

Let $y = a^x$ and take logs to the base e of both sides.

> $\ln a^x = x \ln a$ by third law of logarithms.

$$y = a^x \Rightarrow \ln y = \ln a^x = x \ln a$$

$$x = \frac{1}{\ln a} \ln y$$

$$\frac{dx}{dy} = \frac{1}{\ln a}\frac{1}{y} \Rightarrow \frac{dy}{dx} = \frac{1}{\dfrac{dx}{dy}} = y \ln a = a^x \ln a$$

> **KEY POINT**
>
> If $y = a^x$, then $\dfrac{dy}{dx} = a^x \ln a.$

Parametric functions

> **Core 4**

> **KEY POINT**
>
> If x and y are each expressed in terms of a **parameter** t,
> then $\dfrac{dy}{dx} = \dfrac{dy}{dt} \times \dfrac{dt}{dx}.$ Remember that $\dfrac{dt}{dx} = \dfrac{1}{dx/dt}.$

Example

A curve is defined parametrically by $x = 3t^2$, $y = 4t$.

Find $\dfrac{dy}{dx}$ when $y = 8$

$$x = 3t^2 \Rightarrow \frac{dx}{dt} = 6t \Rightarrow \frac{dt}{dx} = \frac{1}{6t}$$

$$y = 4t \Rightarrow \frac{dy}{dt} = 4$$

$$\therefore \quad \frac{dy}{dx} = \frac{dy}{dt} \times \frac{dt}{dx} = 4 \times \frac{1}{6t} = \frac{2}{3t}$$

> Find the value of t when $y = 8$.

When $y = 8$, $4t = 8$, i.e. $t = 2$; when $t = 2$, $\dfrac{dy}{dx} = \dfrac{2}{3t} = \dfrac{2}{6} = \dfrac{1}{3}.$

Implicit functions

Core 4

> To find $\dfrac{dy}{dx}$ when an equation in x and y is given **implicitly**, differentiate each term with respect to x, remembering that $\dfrac{d}{dx}(f(y)) = \dfrac{d}{dy}f(y) \times \dfrac{dy}{dx}$.

Example

The point $P(2, 6)$ lies on the circle $x^2 + y^2 + 2x - 4y - 20 = 0$.

(a) Show that $\dfrac{dy}{dx} = \dfrac{x+1}{2-y}$

(b) Calculate the gradient of the tangent at P.

(a) $2x + 2y\dfrac{dy}{dx} + 2 - 4\dfrac{dy}{dx} + 0 = 0$

> Differentiate term by term with respect to x
> $\dfrac{d}{dx}(y) = \dfrac{dy}{dx}$.

$x + y\dfrac{dy}{dx} + 1 - 2\dfrac{dy}{dx} = 0$

> Divide throughout by 2 to simplify the equation.

$x + 1 = 2\dfrac{dy}{dx} - y\dfrac{dy}{dx}$

> Get the terms in $\dfrac{dy}{dx}$ on one side of the equation and all the other terms on the other side.

$\Rightarrow x + 1 = \dfrac{dy}{dx}(2 - y)$

> Take out $\dfrac{dy}{dx}$ as a factor.

$\therefore \dfrac{dy}{dx} = \dfrac{x+1}{2-y}$

(b) At P, $x = 2$ and $y = 6$,

so $\dfrac{dy}{dx} = \dfrac{3}{-4} = -\dfrac{3}{4}$.

> The value of $\dfrac{dy}{dx}$ when $x = 2$ gives the gradient of the tangent at $(2, 6)$

The gradient of the tangent at P is $-\dfrac{3}{4}$.

Progress check

Core 3

1 Differentiate with respect to x:

   (a) $y = 2 \sin 6x$ (b) $y = 4 \cos^3 x$ (c) $y = 2 \tan 3x$

   (d) $y = e^{\sin x}$ (e) $y = \sin(x^3)$

Core 3

2 If $y = \sin^2 x$, show that $\dfrac{dy}{dx} = \sin 2x$.

> Use a trigonometric identity.

Core 3

3 Differentiate with respect to x:

   (a) $\ln(x^2 + 3)$ (b) $\ln(2x - 4)^5$ (c) $\ln(\cos 2x)$ (d) $\ln(6x - 1)$

Core 3

4 Find $\dfrac{dy}{dx}$, simplifying your answers.

   (a) $y = x^2 \cos 3x$ (b) $y = e^{2x} \sin x$ (c) $y = x \ln x$ (d) $y = (x + 1)(2x - 3)^4$

Core 3

5 Find the equation of the tangent to the curve $y = x\sqrt{1 + 2x}$ at the point $(4, 12)$.

Core 3

6 Use the quotient rule to find $\dfrac{dy}{dx}$, simplifying your answers where necessary.

   (a) $y = \dfrac{\sin 2x}{x}$ (b) $y = \dfrac{2x - 1}{3x + 4}$ (c) $y = \dfrac{\ln(x + 2)}{x + 2}$

Core 4

7 A curve is defined parametrically by $x = t^2$, $y = t^3$.

   (a) Find $\dfrac{dy}{dx}$ when $t = 2$.

   (b) Find the Cartesian equation of the curve.

Core 4

8 A curve has equation $x^2 + xy^2 - 5 = 0$.

   (a) Find the value of $\dfrac{dy}{dx}$ at $(1, 2)$.

   (b) Find the equation of the normal when $x = 1$, in the form $ax + by + c = 0$.

8 (a) -1.5 (b) $2x - 3y + 4 = 0$

7 (a) 3 (b) $x^3 = y^2$

6 (a) $\dfrac{2x \cos 2x - \sin 2x}{x^2}$ (b) $\dfrac{11}{(3x + 4)^2}$ (c) $\dfrac{1 - \ln(x + 2)}{(x + 2)^2}$

5 $3y = 13x - 16$

4 (a) $x(2 \cos 3x - 3x \sin 3x)$ (b) $e^{2x}(\cos x + 2 \sin x)$ (c) $1 + \ln x$ (d) $5(2x + 1)(2x - 3)^3$

3 (a) $\dfrac{2x}{x^2 + 3}$ (b) $\dfrac{5}{x - 2}$ (c) $-2 \tan 2x$ (d) $\dfrac{6}{6x - 1}$

1 (a) $12 \cos 6x$ (b) $-12 \sin x \cos^2 x$ (c) $6 \sec^2 3x$ (d) $\cos x e^{\sin x}$ (e) $3x^2 \cos(x^3)$

1.6 Integration

After studying this section you should be able to:

- *integrate e^x and simple variations of this function with respect to x*
- *integrate trigonometrical functions*
- *recognise an integral that leads to a logarithmic function*
- *integrate using partial fractions*
- *integrate using a substitution*
- *use integration by parts to integrate a product*
- *find the area under a curve defined parametrically*
- *use integration to find a volume of revolution*
- *solve first order differential equations when the variables can be separated*
- *understand exponential growth and decay*

LEARNING SUMMARY

The exponential function e^x

Core 4

The fact that differentiation leaves the exponential function unchanged was stated on page 32. Since integration is the reverse process of differentiation it follows that

$$\int e^x \, dx = e^x + c \quad \text{and} \quad \int a e^x \, dx = a \int e^x \, dx = \alpha e^x + c.$$

KEY POINT

An extension of this result, using the reverse of the chain rule, is

$$\int e^{ax+b} \, dx = \frac{1}{a} e^{ax+b} + c.$$

KEY POINT

Examples

1 $\quad \displaystyle\int 4e^x \, dx = 4e^x + c$

2 $\quad \displaystyle\int \tfrac{1}{2} e^{3x-1} \, dx = \tfrac{1}{6} e^{3x-1} + c$

Trigonometric functions

Core 4

The following integrals are obtained by applying the reverse process of differentiating the three main trigonometric functions (page 34).

You should learn these and practise using them.

$$\int \cos x \, dx = \sin x + c \qquad \int \cos(ax+b) \, dx = \frac{1}{a} \sin(ax+b) + c$$

$$\int \sin x \, dx = -\cos x + c \qquad \int \sin(ax+b) \, dx = -\frac{1}{a} \cos(ax+b) + c$$

$$\int \sec^2 x \, dx = \tan x + c \qquad \int \sec^2(ax+b) \, dx = \frac{1}{a} \tan(ax+b) + c$$

Example

It is a good idea to check your answer by differentiating.

$$\int (4 \cos 2x - \sin 2x) \, dx = 2 \sin 2x + \tfrac{1}{2} \cos 2x + c$$

Example

When integrating, the angle must be in radians.

Learn the exact values of the trigonometric ratios of $\dfrac{\pi}{6}, \dfrac{\pi}{4}, \dfrac{\pi}{3}$ (30°, 45°, 60°).

$$\int_{\frac{\pi}{3}}^{\frac{\pi}{2}} \sec^2(\tfrac{1}{2}x)\,dx = \left[\frac{1}{\frac{1}{2}}\tan(\tfrac{1}{2}x)\right]_{\frac{\pi}{3}}^{\frac{\pi}{2}}$$

$$= 2\left[\tan(\tfrac{1}{2}x)\right]_{\frac{\pi}{3}}^{\frac{\pi}{2}}$$

$$= 2\left(\tan\frac{\pi}{4} - \tan\frac{\pi}{6}\right)$$

$$= 2\left(1 - \frac{1}{\sqrt{3}}\right)$$

These results are very useful and can be recognised easily:

$$\frac{d}{dx}(\sin^{n+1}x) = (n+1)\sin^n x \cos x \quad \Rightarrow \quad \int \sin^n x \cos x \, dx = \frac{1}{n+1}\sin^{n+1}x + c$$

$$\frac{d}{dx}(\cos^{n+1}x) = -(n+1)\cos^n x \sin x \quad \Rightarrow \quad \int \cos^n x \sin x \, dx = -\frac{1}{n+1}\cos^{n+1}x + c$$

Examples

$$\int \sin^5 x \cos x \, dx = \tfrac{1}{6}\sin^6 x + c$$

$$\int \sin x \cos^3 x \, dx = -\tfrac{1}{4}\cos^4 x + c$$

Odd powers of sin x and cos x

Example

Use $\cos^2 x = 1 - \sin^2 x$.

Recognise the integral of $\cos x \sin^2 x$.

Evaluate $\displaystyle\int_0^{\frac{\pi}{6}} \cos^3 x \, dx$

$$\int_0^{\frac{\pi}{6}} \cos^3 x \, dx = \int_0^{\frac{\pi}{6}} \cos x (\cos^2 x)\,dx$$

$$= \int_0^{\frac{\pi}{6}} \cos x (1 - \sin^2 x)\,dx$$

$$= \int_0^{\frac{\pi}{6}} (\cos x - \cos x \sin^2 x)\,dx$$

$$= \left[\sin x - \tfrac{1}{3}\sin^3 x\right]_0^{\frac{\pi}{6}}$$

$$= \tfrac{1}{2} - \tfrac{1}{24} = \tfrac{11}{24}$$

Even powers of cos x and sin x

To integrate even powers of sin x and cos x, use the double angle identities for cos $2x$.

$$\cos 2x = 2\cos^2 x - 1 \quad \Rightarrow \quad \cos^2 x = \tfrac{1}{2}(1 + \cos 2x)$$

$$\cos 2x = 1 - 2\sin^2 x \quad \Rightarrow \quad \sin^2 x = \tfrac{1}{2}(1 - \cos 2x)$$

Example

$$\int \cos^2 x \, dx = \tfrac{1}{2} \int (1 + \cos 2x) dx$$

$$= \tfrac{1}{2}(x + \tfrac{1}{2} \sin 2x) + c$$

To integrate $\cos^4 x$, write it as $(\cos^2 x)^2 = (\tfrac{1}{2}(1 + \cos 2x))^2$.

The reciprocal function $\dfrac{1}{x}$

Core 4

The result $\int x^n \, dx = \dfrac{x^{n+1}}{n+1} + c$ holds for any value of n other than $n = -1$.

In this special case, the result takes a different form:

> Remember that $x^{-1} = \dfrac{1}{x}$.

> **KEY POINT**
> $$\int \frac{1}{x} \, dx = \ln|x| + c.$$

This extends to $\displaystyle\int \frac{k}{x} \, dx = k \int \frac{1}{x} \, dx = k \ln|x| + c,$

and to $\displaystyle\int \frac{1}{kx} \, dx = \frac{1}{k} \int \frac{1}{x} \, dx = \frac{1}{k} \ln|x| + c.$ (where k is a constant)

For example, $\displaystyle\int \frac{2}{x} \, dx = 2 \ln|x| + c$ and $\displaystyle\int \frac{1}{3x} \, dx = \frac{1}{3} \ln|x| + c.$

Integrals leading to a logarithmic function

Core 4

Remember that $\dfrac{d}{dx} \ln(f(x)) = \dfrac{f'(x)}{f(x)}.$

Applying the process in reverse gives:

> **KEY POINT**
> $$\int \frac{f'(x)}{f(x)} \, dx = \ln|f(x)| + c.$$

Examples

(a) $\displaystyle\int \frac{x^2}{2x^3 - 2} \, dx = \tfrac{1}{6} \ln|2x^3 - 2| + c$

> $\dfrac{d}{dx}(2x^3 - 2) = 6x^2$ so a factor of $\tfrac{1}{6}$ is needed.

(b) $\displaystyle\int \tan x \, dx = \int \frac{\sin x}{\cos x} \, dx$

$$= -\ln|\cos x| + c$$

$$= \ln|\sec x| + c$$

> $-\ln a = \ln(a^{-1}).$

Care must be taken with **definite integrals** of this type.
These can be evaluated between the limits a and b only if $f(x)$ exists for all values between a and b. Note that $f(a)$ and $f(b)$ will have the same sign.

$$\int_a^b \frac{f'(x)}{f(x)} \, dx = \left[\ln|f(x)| \right]_a^b = (\ln|f(b)| - \ln|f(a)|)$$

Key points from AS

• **Area under curve**
  Revise AS page 53

Example

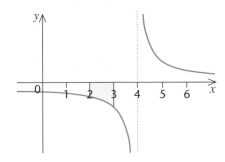

The diagram shows the curve $y = \dfrac{1}{x-4}$.

Find the area between the curve, the x-axis, and the lines $x = 2$ and $x = 3$.

$$\int_a^b y \, dx = \int_2^3 \frac{1}{x-4} \, dx$$

$$= \left[\ln|x-4| \right]_2^3$$

$\ln|-2| = \ln 2$
$\ln 1 = 0$.

$$= \ln|-1| - \ln|-2|$$

$$= \ln 1 - \ln 2$$

$$= -\ln 2 \ (= -0.69 \ldots)$$

Note that the formula gives a negative value, confirming that the area is below the x-axis.

Shaded area $= \ln 2 \ (= 0.69 \ldots)$ square units.

$f(3)$ and $f(5)$ do not have the same sign.

Note that $\displaystyle\int_3^5 \frac{1}{x-4} \, dx$ cannot be evaluated, as the curve is undefined at $x = 4$.

Partial fractions

Core 4

Some rational functions can be integrated by expressing them in partial fraction form.

Example

Find $\displaystyle\int \frac{x+7}{(x-2)(x+1)} \, dx$, simplifying your answer.

Decompose into partial fractions (see page 15).

$$\frac{x+7}{(x-2)(x+1)} \equiv \frac{3}{x-2} - \frac{2}{x+1}$$

The working for this example is shown on page 15.

$$\Rightarrow \int \frac{x+7}{(x-2)(x+1)} \, dx = \int \left(\frac{3}{x-2} - \frac{2}{x+1} \right) dx$$

$$= 3\ln(x-2) - 2\ln(x+1) + c$$

$$= \ln k \frac{(x-2)^3}{(x+1)^2}$$

$\log a - \log b = \log\left(\dfrac{a}{b}\right)$

$\log a^n = n \log a$.

Key points from AS

• **Logarithms**
  Revise AS page 49

Note: Sometimes the constant c is incorporated into the log function by letting $c = \ln k$.

Substitution

Sometimes an integral is made easier by using a substitution.

This transforms the integral with respect to one variable, say x, into an integral with respect to a related variable, say u.

$$\int f(x)\,dx = \int f(x)\,\frac{dx}{du}\,du.$$

KEY POINT

This method is known as the method of **substitution** or **change of variable**.

> This could also be done by using the substitution $u = 2x + 1$.

Example

Find $\int x\sqrt{2x + 1}\ dx$, using the substitution $u = \sqrt{2x + 1}$.

$$\int x\sqrt{2x + 1}\ dx = \int x\sqrt{2x + 1}\,\frac{dx}{du}\,du$$

Side working

$$= \int \tfrac{1}{2}(u^2 - 1)u \times u\,du$$

$$u = \sqrt{2x + 1}$$

> Show the side working as part of your answer.

$$= \frac{1}{2}\int (u^4 - u^2)\,du$$

$$u^2 = 2x + 1 \implies x = \tfrac{1}{2}(u^2 - 1)$$

> When simplifying, take out factors as soon as possible.

$$= \frac{1}{2}\left(\frac{u^5}{5} - \frac{u^3}{3}\right) + c$$

$$\frac{dx}{du} = u$$

$$= \frac{u^3}{2}\left(\frac{u^2}{5} - \frac{1}{3}\right) + c$$

> Write the answer in terms of x, simplifying as much as possible.

$$= \tfrac{1}{30}u^3(3u^2 - 5) + c$$

$$3u^2 - 5 = 3(2x + 1) - 5$$

$$= \tfrac{1}{30}(2x + 1)^{\frac{3}{2}}(6x - 2) + c$$

$$= 6x - 2$$

$$= \tfrac{1}{15}(2x + 1)^{\frac{3}{2}}(3x - 1) + c$$

$$= 2(3x - 1)$$

When evaluating definite integrals using the method of substitution, change the x limits to u limits and substitute them into the working as soon as possible. There is no need to tidy up the expression first.

Sometimes it is possible to bypass the method of substitution by **recognising** an integral as the derivative of a particular function.

Examples

> This type of integral can always be done by using a substitution.

(a) $\int (2x + 1)^5\ dx = \tfrac{1}{12}(2x + 1)^6 + c$

Recognise that
$$\frac{d}{dx}(2x + 1)^6 = 12(2x + 1)^5.$$

(b) $\int x\sqrt{2x^2 + 1}\ dx = \tfrac{1}{6}(2x^2 + 1)^{\frac{3}{2}} + c$

Recognise that
$$\frac{d}{dx}(2x^2 + 1)^{\frac{3}{2}} = 6x(2x^2 + 1)^{\frac{1}{2}}.$$

General result:

$$\int f'(x)[f(x)]^n\ dx = \frac{1}{n + 1}[f(x)]^{n+1} + c$$

Integration by parts

Core 4

It is sometimes possible to integrate a product using **integration by parts**.

If u and v are functions of x, then $\displaystyle\int u\frac{dv}{dx}dx = uv - \int v\frac{du}{dx}dx$.

(This formula is given in the exam.)

Before using integration by parts, check whether the function can be simplified, or whether the integral can be recognised directly or done using a substitution.

Example

Find $\displaystyle\int 2x\sin x\, dx$.

Let $u = 2x$, then $\dfrac{du}{dx} = 2$

Let $\dfrac{dv}{dx} = \sin x$ then $v = -\cos x$

Integration by parts can only be used if v can be found from $\dfrac{dv}{dx}$ and it is possible to find $\displaystyle\int v\frac{du}{dx}dx$.

$\therefore \displaystyle\int 2x\sin x\, dx = 2x(-\cos x) - \int(-\cos x \times 2)dx$

$\qquad = -2x\cos x + \displaystyle\int 2\cos x\, dx$

$\qquad = -2x\cos x + 2\sin x + c$

Sometimes the integration by parts process has to be carried out more than once. This happens when integrating expressions such as $x^2 e^x$ or $x^2\cos x$.

Definite integrals

$$\int_a^b u\frac{dv}{dx}dx = \left[uv\right]_a^b - \int_a^b v\frac{du}{dx}dx$$

Example

$\displaystyle\int_0^1 xe^{2x}dx = [x\tfrac{1}{2}e^{2x}]_0^1 - \int_0^1 \tfrac{1}{2}e^{2x}dx$

$\qquad = \tfrac{1}{2}e^2 - [\tfrac{1}{4}e^{2x}]_0^1$

$\qquad = \tfrac{1}{2}e^2 - (\tfrac{1}{4}e^2 - \tfrac{1}{4}) = \tfrac{1}{4}e^2 + \tfrac{1}{4}$

Working:

Let $u = x$, then $\dfrac{du}{dx} = 1$

Let $\dfrac{dv}{dx} = e^{2x}$, then $v = \tfrac{1}{2}e^{2x}$

Some functions containing $\ln x$ can be integrated using integration by parts. A useful strategy is to take $\ln x$ as u.

Often a polynomial function is taken as u because it becomes of lower degree when differentiated. Integrals involving $\ln x$ are an exception, since $\ln x$ is not easy to integrate, but can be differentiated.

Example

$\displaystyle\int x^2\ln x\, dx = \int(\ln x \times x^2)dx$

$\qquad = \ln x \times \left(\dfrac{x^3}{3}\right) - \displaystyle\int\left(\dfrac{x^3}{3}\times\dfrac{1}{x}\right)dx$

$\qquad = \tfrac{1}{3}x^3\ln x - \tfrac{1}{3}\displaystyle\int x^2\, dx$

$\qquad = \tfrac{1}{3}x^3\ln x - \tfrac{1}{9}x^3 + c$

Working:

Let $u = \ln x$, then $\dfrac{du}{dx} = \dfrac{1}{x}$

Let $\dfrac{dv}{dx} = x^2$, then $v = \dfrac{x^3}{3}$

This method is particularly useful when integrating $\ln x$.
Think of it as integrating the product $\ln x \times 1$ as follows:

$$\int (\ln x \times 1) \, dx = \ln x \times x - \int \left(x \times \frac{1}{x} \right) dx$$

$$= x \ln x - \int 1 \, dx$$

$$= x \ln x - x + c$$

Working:

Let $u = \ln x$, then $\dfrac{du}{dx} = \dfrac{1}{x}$

Let $\dfrac{dv}{dx} = 1$, then $v = x$

Parametric integration

Core 4

When a curve is defined in terms of a parameter such that $x = f(t)$ and $y = g(t)$ then the formula for the area under a curve can be adapted as follows:

$$A = \int_{x_1}^{x_2} y \, dx = \int_{t_1}^{t_2} g(t) \frac{dx}{dt} \, dt.$$

KEY POINT

Example
Using parametric integration, find the area of the ellipse defined by $x = 2 \cos \theta$, $y = 3 \sin \theta$ $(0 \leqslant \theta \leqslant 2\pi)$.

The sign of the area given by the formula depends on the direction in which the curve is being traced out as the parameter increases: positive for clockwise, negative for anticlockwise.

$$A = \int_0^{2\pi} y \frac{dx}{d\theta} \, d\theta$$

$$= \int_0^{2\pi} 3 \sin \theta (-2 \sin \theta) d\theta$$

$$= -6 \int_0^{2\pi} \sin^2 \theta \, d\theta$$

$$= -6 \int_0^{2\pi} \tfrac{1}{2} (1 - \cos 2\theta) d\theta$$

$$= -3 \left[\theta - \tfrac{1}{2} \sin 2\theta \right]_0^{2\pi}$$

$$= -3(2\pi - 0 - (0 - 0))$$

$$= -6\pi$$

In general, area of ellipse defined by $x = a \cos \theta$, $y = b \sin \theta$ is $\pi a b$.

This curve is traced out in an anticlockwise direction

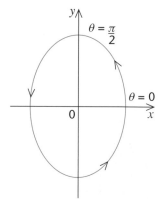

The area of the ellipse is 6π square units.
The shaded region R is bounded by the curve $y = f(x)$, the x-axis and the lines $x = a$ and $x = b$.

Volume of revolution about the x-axis

Core 4

Rotating R completely about the x-axis forms a solid figure. The volume of this figure is called the volume of revolution and is given by

$$V = \int_a^b \pi y^2 \, dx$$

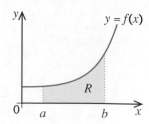

Example

Find the volume of the solid generated when the following functions are rotated about the x-axis:

(a) $y = 2x^2 - 4$ from $x = 3$ to $x = 5$

(b) $x = t^2 + 1$, $y = t$ from $t = 2$ to $t = 3$

(a) $V = \displaystyle\int_3^5 \pi\,(2x^2 - 4)^2 \,\mathrm{d}x = \pi \int_3^5 4x^4 - 16x^2 + 16 \,\mathrm{d}x = \pi \left[\frac{4x^5}{5} - \frac{16x^3}{3} + 16x\right]_3^5 = 5702$

(b) $V = \displaystyle\int \pi\, t^2 \,\mathrm{d}x = \pi \int_2^3 t^2 \frac{\mathrm{d}x}{\mathrm{d}t} \,\mathrm{d}t = \pi \int_2^3 t^2(2t)\,\mathrm{d}t = \pi \left[\frac{t^4}{2}\right]_2^3 = \frac{65\pi}{2}$

Differential equations

Core 4

A **first order differential equation** in x and y contains only the first differential coefficient.

Examples $\dfrac{\mathrm{d}y}{\mathrm{d}x} = -2x$, $\dfrac{\mathrm{d}y}{\mathrm{d}x} = \dfrac{x}{y}$, $y\dfrac{\mathrm{d}y}{\mathrm{d}x} = 3$

If $\dfrac{\mathrm{d}y}{\mathrm{d}x} = -2x$, then integrating with respect to x gives $y = -x^2 + c$.

> The general solution contains an arbitrary integration constant.

This is the **general solution** of the differential equation and it can be illustrated by a family of curves.

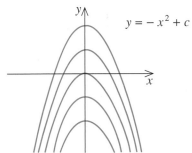

A **particular solution** is a specific member of the family and it can be found from additional information.

For example, if the curve $y = -x^2 + c$ goes through the point $(2, 0)$, then $0 = -(2)^2 + c \Rightarrow c = 4$. The particular solution is $y = -x^2 + 4$.

> The value of c is found to give a particular solution.

Separating the variables

> **KEY POINT**
>
> A differential equation that can be written in the form $f(y)\dfrac{\mathrm{d}y}{\mathrm{d}x} = g(x)$
>
> can be solved by **separating the variables**, where $\displaystyle\int f(y)\mathrm{d}y = \int g(x)\mathrm{d}x$.

Example

Solve (a) $\dfrac{\mathrm{d}y}{\mathrm{d}x} = \dfrac{x}{y}$ (b) $\dfrac{\mathrm{d}y}{\mathrm{d}x} = y$

(a) $\dfrac{\mathrm{d}y}{\mathrm{d}x} = \dfrac{x}{y}$

> Separate the variables to get the format $\int \ldots \mathrm{d}y = \int \ldots \mathrm{d}x$.

$\displaystyle\int y \,\mathrm{d}y = \int x \,\mathrm{d}x \Rightarrow \tfrac{1}{2}y^2 = \tfrac{1}{2}x^2 + c$

(b) $\quad \dfrac{dy}{dx} = y$

This is a special case, when $g(x) = 1$.

$$\dfrac{1}{y}\dfrac{dy}{dx} = 1$$

$$\int \dfrac{1}{y}\,dy = \int 1\,dx$$

The answer can be left like this.

$$\ln y = x + c$$

$$y = e^{x+c} = e^x \times e^c = Ae^x \quad (\text{where } A = e^c)$$

The letter A is often used here, but it could be any letter (other than e, x or y, of course!).

$$\text{So } \dfrac{dy}{dx} = y \Longrightarrow y = Ae^x$$

Exponential growth and decay

Core 4

You may have to form a differential equation from given information. Often this models exponential growth or exponential decay.

Exponential growth   **Exponential decay**

     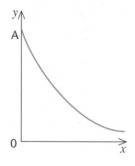

In exponential decay, the fact that y is decreasing with respect to x is usually shown by the negative, keeping $k > 0$.

$$\dfrac{dy}{dx} = ky \Longrightarrow y = Ae^{kx} \quad (k > 0) \qquad \dfrac{dy}{dx} = -ky \Longrightarrow y = Ae^{-kx} \quad (k > 0)$$

An example of exponential growth is the growth of a population where

p_0 is the initial population.

$$\dfrac{dp}{dt} = kt \Longrightarrow p = p_0 e^{-kt}$$

Examples of exponential decay are:

m_0 is the initial mass.

(a) the disintegration of radioactive materials, $\dfrac{dm}{dt} = -km \Longrightarrow m = m_0 e^{-kt}$

T_s is temperature of surroundings, T_0 is initial temperature of the object.

(b) Newton's Law of Cooling, $\dfrac{dT}{dt} = -k(T - T_s) \Longrightarrow T - T_s = (T_0 - T_s)e^{-kt}$

Progress check

Core 4

1 Integrate with respect to x.

  (a) $\dfrac{3}{x}$ (b) $\dfrac{x+3}{x^2}$ (c) $3e^x - \dfrac{1}{x}$ (d) $3e^{4-x}$

Core 4

2 The region bounded by the curve $y = e^{\frac{x}{2}}$, the lines $x = 1$ and $x = 3$ and the x-axis is rotated completely about the x-axis. Find the volume of revolution.

Core 4

3 (a) $\displaystyle\int 4\sin 2x\, dx$ (b) $\displaystyle\int \sec^2 2x\, dx$ (c) $\displaystyle\int \cos\left(\tfrac{1}{2}x\right) dx$

  (d) $\displaystyle\int 4\sin^2 x\, dx$ (e) $\displaystyle\int \cos^2 2x\, dx$ (f) Evaluate $\displaystyle\int_0^{\frac{1}{6}\pi} \sin^4 x \cos x\, dx$

Core 4

4 Find (a) $\displaystyle\int \dfrac{1}{2x+5}\, dx$ (b) $\displaystyle\int \cot x\, dx$ (c) $\displaystyle\int \dfrac{3x+2}{x^2-x-12}\, dx$

> Split (c) into partial fractions first.

Core 4

5 Using the substitution $u = 2 + e^{2x}$, or otherwise, find $\displaystyle\int \dfrac{e^{2x}}{2+e^{2x}}\, dx$

> Remember that $\dfrac{dx}{du} = \dfrac{1}{du/dx}$.

Core 4

6 (a) Find (i) $\displaystyle\int xe^{-x}\, dx$ (ii) $\displaystyle\int x^3 \ln x\, dx$ (b) Evaluate $\displaystyle\int_0^{\frac{1}{2}\pi} x\cos x\, dx$

Core 4

7 Solve the differential equation $\dfrac{dy}{dx} = 2xy$, given that $y = 3$ when $x = 0$.

---

7 $y = 3e^{x^2}$

6 (a) (i) $-e^{-x}(x+1)+c$ (ii) $\frac{1}{16}x^4(4\ln x - 1)+c$ (b) $\frac{1}{2}\pi - 1$

5 $\frac{1}{2}\ln(2+e^{2x})+c$

4 (a) $\frac{1}{2}\ln(2x+5)+c$ (b) $\ln(\sin x)+c$ (c) $\ln(x+3)(x-4)^2+c$ or $\ln A(x+3)(x-4)^2$

3 (a) $-2\cos 2x+c$ (b) $\frac{1}{2}\tan 2x+c$ (c) $2\sin\left(\frac{x}{2}\right)+c$
    (d) $2x - \sin 2x+c$ (e) $\frac{x}{2}+\frac{1}{8}\sin 4x+c$ (f) $\frac{1}{160}$

2 $\displaystyle\int_1^3 \pi e^x\, dx = \pi[e^x]_1^3 = \pi(e^3 - e)$.

1 (a) $3\ln|x|+c$ (b) $\ln|x| - \frac{3}{x}+c$ (c) $3e^x - \ln|x|+c$ (d) $-3e^{4-x}+c$

1.7 Numerical methods

After studying this section you should be able to:

- locate an interval containing a root of an equation by finding a change of sign
- use simple iteration to solve equations
- find the approximate value of an integral using the trapezium rule

Change of sign

Core 3

If there is an interval from $x = a$ to $x = b$ in which the graph of a function has no breaks then the function is said to be **continuous** on the interval.

> All polynomial functions are continuous everywhere.

For example, the graph of $y = x^2 - 3$ is continuous everywhere.

However, the graph of $y = \dfrac{1}{x-3}$ is not continuous on any interval that contains the value $x = 3$.

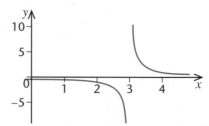

> **KEY POINT**
>
> If $f(x)$ is continuous between $x = a$ and $x = b$ and if $f(a)$ and $f(b)$ have different signs, then a root of $f(x) = 0$ lies in the interval from a to b.

Example Given that $f(x) = e^x - 10x$, show that the equation $f(x) = 0$ has a root between 3 and 4.

$f(3) = -9.914... < 0$

$f(4) = 14.598... > 0$.

The change of sign shows that $f(x) = 0$ has a root between 3 and 4.

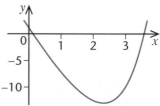

You can continue this process, called a **decimal search**, to trap the root in a smaller and smaller interval. The table gives the values of the function in steps of 0.1.

> Most graphic calculators can tabulate values of a function and this gives a very efficient way to obtain the information.

x	3.1	3.2	3.3	3.4	3.5	3.6	3.7	3.8	3.9
$f(x)$	−8.802	−7.467	−5.887	−4.035	−1.884	0.5982	3.4473	6.7011	10.402

This shows that the root lies between 3.5 and 3.6
At the mid-point $f(3.55) = -0.6866... < 0$ and so the root lies between 3.55 and 3.6

This is a good way to establish the value of a root to a particular level of accuracy.

| 3.5 | 3.55 | 3.6 |

It follows that the value of the root is 3.6 to 1 d.p.

Iteration

Core 3

To solve an equation by **iteration** you start with some approximation to a root and improve its accuracy by substituting it into a formula. You can then repeat the process until you have the desired level of accuracy.

There are different methods for producing an iterative formula. You only need to know about simple iteration for this component.

Using **simple iteration**, the first step is to rearrange the equation to express x as a function of itself. This function defines the iterative formula that you need.

For example, the equation $e^x - 10x = 0$ can be rearranged as

$$e^x = 10x$$
$$\Rightarrow x = \ln(10x).$$

This may now be turned into an iterative formula for finding x.

$$x_{n+1} = \ln(10x_n).$$

Starting with $x_1 = 3$, $x_2 = \ln 30 = 3.401 \ldots$ and so on. This produces a sequence of values:

The values may be found very easily using a calculator's Ans function: Key in 3 = followed by ln(10Ans) = = = to produce the sequence. (Depending on the make of your calculator you may need to use EXE or ENTER in place of = .)

$x_2 = 3.401 \ldots$	$x_6 = 3.5760 \ldots$
$x_3 = 3.526 \ldots$	$x_7 = 3.5768 \ldots$
$x_4 = 3.563 \ldots$	$x_8 = 3.5770 \ldots$
$x_5 = 3.573 \ldots$	$x_9 = 3.5771 \ldots$

The calculator display soon settles on 3.577152064 The value of the root is $x = 3.577$ to 4 s.f.

Start from a point on the x-axis, move up to the curve and across to $y = x$. Then move up to the curve and across to $y = x$ again. Continue in the same way. Each of these stages corresponds to one iteration of the formula.

The diagram shows how the process converges from starting point on either side of the root.

A different rearrangement of the original equation gives $x = \dfrac{e^x}{10}$ and so the corresponding iterative formula is $x_{n+1} = \dfrac{e^{x_n}}{10}$.

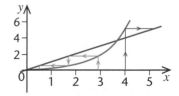

You may need to try several rearrangements in order to converge to a particular root of an equation by this method.

Starting with $x = 3$, this rearrangement fails to converge to the root between 3 and 4 but it does converge to the other root of the equation.

Notice how the movement is away from the upper root this time. If the starting value is smaller than the upper root then the process converges to the lower root. A starting value above the upper root fails to converge to either root.

The value of this root is $x = 0.1118$ to 4 s.f.

In some exam questions you may be given the iteration formula to start with.

Example Starting with $x_1 = 1$, use the formula $x_{n+1} = \dfrac{x_n^2 - 5x_n}{10} - 3$ to find $x_2, x_3, \ldots, x_8$ and describe the long term behaviour of the sequence.

Try the key sequence:
1 =
(Ans² – 5Ans)/10 – 3
=
=
=
=

$x_2 = -3.4$

$x_3 = -0.144$

$x_4 = -2.9259264$

$x_5 = -0.6809322$

$x_6 = -2.6131669$

$x_7 = -1.0105523$

$x_8 = -2.3926032$

The sequence oscillates but converges to -1.787 to 4 s.f. which is the lower root of the equation

$$x = \frac{x^3 - 5x}{10} - 3,$$

which simplifies to

$$x^2 - 15x - 30 = 0.$$

The oscillations correspond to a cobweb pattern on the diagram.

Numerical integration

Core 4

This topic was also covered in Core 2. See Revise AS page 54. You will be expected to apply the trapezium rule to functions covered in Core 3 and Core 4. You may be required to integrate to find the **exact** value of the area and compare this with the approximation.

The trapezium rule

The value of $A = \int_a^b y\,\mathrm{d}x$ represents the area under the graph of $y = f(x)$ between $x = a$ and $x = b$. You can find an approximation to this value using the **trapezium rule**. This is especially useful if the function is difficult to integrate.

The area under the curve between a and b may be divided into n strips of equal width h.

It follows that $h = \dfrac{b - a}{n}$.

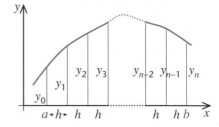

Each strip is approximately a trapezium and so the total area is approximately

$$\tfrac{1}{2}h(y_0 + y_1) + \tfrac{1}{2}h(y_1 + y_2) + \tfrac{1}{2}h(y_2 + y_3) + \ldots + \tfrac{1}{2}h(y_{n-2} + y_{n-1}) + \tfrac{1}{2}h(y_{n-1} + y_n).$$

This simplifies to give the formula known as the trapezium rule:

> **KEY POINT**
>
> This simplifies to give the formula known as the trapezium rule:
>
> $$\int_a^b y\,\mathrm{d}x = \tfrac{1}{2}h\{(y_0 + y_n) + 2(y_1 + y_2 + \ldots + y_{n-1})\} \text{ where } h = \frac{b-a}{n}$$

Example

Use the trapezium rule with five strips to estimate the value of $\int_2^3 \ln x\,\mathrm{d}x$.

It's useful to tabulate the information:

Work to a greater level of accuracy in your table than you intend to give in your final answer.

x	2	2.2	2.4	2.6	2.8	3
$\ln x$	0.69315	0.78846	0.87547	0.95551	1.0296	1.0986
	y_0	y_1	y_2	y_3	y_4	y_5

$$h = \frac{3 - 2}{5} = 0.2$$

Use integration by parts to show that
$$\int_2^3 \ln x\,\mathrm{d}x = \ln \frac{27}{4} - 1$$

$$\int_2^3 \ln x\,\mathrm{d}x \approx 0.1\{(0.69315 + 1.0986) + 2(0.78846 + 0.87547 + 0.95551 + 1.0296)\}$$

$$= 0.9089 \quad = 0.909 \text{ (3 s.f.)}$$

Progress check

Core 3

1 Show that the equation $x^3 - x - 7 = 0$ has a root between 2 and 3.
Use a decimal search to find the value of the root to 2 d.p.

Core 3

2 Use simple iteration to refine the value of the root found in question 1 and
state its value to 4 d.p.

Core 4

3 Use the trapezium rule with five strips to find the value of $\displaystyle\int_{2}^{3} x \ln x \, dx$ to 2 d.p.

Core 3

4 Show that the equation $e^x + 2x^2 - 2 = 0$ has a root between $x = 0.4$ and $x = 0.5$

4 0.46
3 2.31
2 2.0867
1 2.09

1.8 Vectors

After studying this section you should be able to:

- *understand the distinction between vector and scalar quantities*
- *add and subtract vector quantities and multiply by a scalar*
- *use the unit vectors* **i**, **j** *and* **k**
- *find the magnitude and direction of a vector*
- *use scalar products to find the angle between two directions*
- *express equations of lines in vector form*
- *determine whether two lines are parallel, intersect or are skew*

LEARNING SUMMARY

Vector and scalar quantities

Core 4

A **scalar** quantity has size (or **magnitude**) but not direction. **Numbers** are scalars and some other important examples are **distance**, **speed**, **mass** and **time**.

A **vector** quantity has both size and **direction**. For example, distance in a specified direction is called **displacement**.

The diagram shows a **directed line segment**. It has size (in this case, length) and direction so it is a vector.

The diagram gives a useful way to represent *any* vector quantity and may also be used to represent addition and subtraction of vectors and multiplication of a vector by a scalar.

In a textbook, vectors are usually labelled with lower case letters in **bold** print.
When hand-written, these letters should be underlined e.g. *a*.

Addition and subtraction of vectors

Core 4

This diagram shows three vectors **a**, **b** and **c** such that **c** = **a** + **b**.

The vectors **a** and **b** follow on from each other and then **c** joins the start of vector **a** to the end of vector **b**.

Check that **a** + **b** = **b** + **a**.

The vector **c** is called the **resultant** of **a** and **b**.

On a vector diagram, −**q** has the opposite sense of direction to **q**.
Notice that **p** − **q** is represented as **p** + (−**q**).

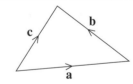

This diagram shows the same information in a different way.

Following the route in the opposite direction to **q** is the same as adding −**q** or subtracting **q**.

Scalar multiplication

Core 4

2p is parallel to **p** and has the same sense of direction but is twice as long.

−3p is parallel to **p** but has the opposite sense of direction and is three times as long.

Component form

Core 4

When working in two dimensions, it is often very useful to express a vector in terms of two special vectors **i** and **j**. These are **unit vectors** at right-angles to each other. A vector **r** written as **r** = a**i** + b**j** is said to have **components** a**i** and b**j**.

A unit vector is a vector of magnitude 1 unit.

Examples

r = **i** + 2**j**

r = 3**i** − **j**

For work involving the Cartesian coordinate system, **i** and **j** are taken to be in the positive directions of the x- and y-axes respectively.

In three dimensions, a third vector **k** is used to represent a unit vector in the positive direction of the z-axis.

The **position vector** of a point P is the vector $\overrightarrow{OP}$ where O is the origin.

If P has coordinates (a, b) then its position vector is given by **r** = a**i** + b**j**.

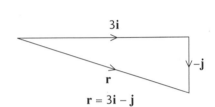

In three dimensions, a point with coordinates (a, b, c) would have position vector **r** = a**i** + b**j** + c**k**.

Adding and subtracting vectors in component form

Core 4

One advantage of expressing vectors in terms of **i**, **j** and **k** is that addition and subtraction may be done algebraically.

Example
p = 3**i** + 2**j** − **k** and **q** = 5**i** − **j** + 4**k**. Express the following vectors in terms of **i**, **j** and **k**:

(a) **p** + **q**   (b) **p** − **q**   (c) 2**p** − 3**q**.

(a) **p** + **q** = (3**i** + 2**j** − **k**) + (5**i** − **j** + 4**k**) = 8**i** + **j** + 3**k**

(b) **p** − **q** = (3**i** + 2**j** − **k**) − (5**i** − **j** + 4**k**) = − 2**i** + 3**j** − 5**k**

(c) 2**p** − 3**q** = 2(3**i** + 2**j** − **k**) − 3(5**i** − **j** + 4**k**)
$\qquad$ = 6**i** + 4**j** − 2**k** − 15**i** + 3**j** − 12**k** = −9**i** + 7**j** − 14**k**.

Finding the magnitude of a vector

Core 4

> **KEY POINT**
>
> If $\mathbf{r} = a\mathbf{i} + b\mathbf{j} + c\mathbf{k}$, the magnitude (length) of $\mathbf{r}$ is given by
> $$|\mathbf{r}| = \sqrt{a^2 + b^2 + c^2}$$

For example,
If $\mathbf{r} = 3\mathbf{i} - 4\mathbf{j} + 12\mathbf{k}$, then $|\mathbf{r}| = \sqrt{3^2 + (-4)^2 + 12^2} = 13$.

Scalar product

Core 4

If $\mathbf{a} = x_1\mathbf{i} + y_1\mathbf{j} + z_1\mathbf{k}$, then $|\mathbf{a}| = \sqrt{x_1^2 + y_1^2 + z_1^2}$ where $|\mathbf{a}|$ is the length of $\mathbf{a}$.

If $\mathbf{b} = x_2\mathbf{i} + y_2\mathbf{j} + z_2\mathbf{k}$, then $|\mathbf{b}| = \sqrt{x_2^2 + y_2^2 + z_2^2}$.

> **KEY POINT**
>
> If θ is the angle between $\mathbf{a}$ and $\mathbf{b}$,
> the **scalar (or dot) product** of $\mathbf{a}$ and $\mathbf{b}$ is $\mathbf{a}.\mathbf{b}$
> where $\mathbf{a}.\mathbf{b} = |\mathbf{a}||\mathbf{b}|\cos\theta$.

This leads to the useful results:
$\mathbf{i}.\mathbf{i} = 1$, $\mathbf{j}.\mathbf{j} = 1$, $\mathbf{k}.\mathbf{k} = 1$
$\mathbf{i}.\mathbf{j} = \mathbf{j}.\mathbf{k} = \mathbf{k}.\mathbf{i} = 0$.

If two vectors are **parallel**, $\theta = 0$ and $\mathbf{a}.\mathbf{b} = |\mathbf{a}||\mathbf{b}|$.
If two vectors are **perpendicular**, $\theta = 90°$ and $\mathbf{a}.\mathbf{b} = 0$.

For $\mathbf{a}$ and $\mathbf{b}$ defined as above:

In column vectors
$$\mathbf{a}.\mathbf{b} = \begin{pmatrix} x_1 \\ y_1 \\ z_1 \end{pmatrix} \cdot \begin{pmatrix} x_2 \\ y_2 \\ z_2 \end{pmatrix}$$
$$= x_1 x_2 + y_1 y_2 + z_1 z_2.$$

$\mathbf{a}.\mathbf{b} = (x_1\mathbf{i} + y_1\mathbf{j} + z_1\mathbf{k}) \cdot (x_2\mathbf{i} + y_2\mathbf{j} + z_2\mathbf{k})$
$\quad = x_1 x_2 + y_1 y_2 + z_1 z_2$

This uses the results for parallel and perpendicular vectors.

Example
Find the angle between $\mathbf{a} = 2\mathbf{i} + \mathbf{j} + 4\mathbf{k}$ and $\mathbf{b} = -3\mathbf{i} + 2\mathbf{j} - \mathbf{k}$.

$\mathbf{a}.\mathbf{b} = 2(-3) + 1(2) + 4(-1) = -8$
$|\mathbf{a}| = \sqrt{2^2 + 1^2 + 4^2} = \sqrt{21}$
$|\mathbf{b}| = \sqrt{(-3)^2 + 2^2 + (-1)^2} = \sqrt{14}$
$\mathbf{a}.\mathbf{b} = |\mathbf{a}||\mathbf{b}|\cos\theta$
$-8 = \sqrt{21}\sqrt{14}\cos\theta$
$\Rightarrow \cos\theta = \dfrac{-8}{\sqrt{21}\sqrt{14}} = -0.4665\ldots$
$\Rightarrow \theta = 117.8°$ (1 d.p.)

In column vectors:
$$\mathbf{a}.\mathbf{b} = \begin{pmatrix} 2 \\ 1 \\ 4 \end{pmatrix} \cdot \begin{pmatrix} -3 \\ 2 \\ -1 \end{pmatrix} = -8.$$

Vector equation of a line

Core 4

> **KEY POINT**
>
> A **vector equation** of a line passing
> through a fixed point A with position
> vector $\mathbf{a}$ and parallel to a vector $\mathbf{b}$ is
> $\mathbf{r} = \mathbf{a} + t\mathbf{b}$, where t is a scalar
> parameter.

t is often used for the parameter, but not exclusively so. Other letters sometimes used include s, μ and λ.

r is the position vector of any point on the line.
b is the **direction vector** of the line.

The equation of the line is not unique. Any other point on the line could be used instead of A and any multiple of **b** could be used for the direction vector.

Example
A vector equation of the line passing through $(4, -5, 1)$, parallel to $3\mathbf{i} + 4\mathbf{j} - 2\mathbf{k}$ is
$\mathbf{r} = 4\mathbf{i} - 5\mathbf{j} + \mathbf{k} + t(3\mathbf{i} + 4\mathbf{j} - 2\mathbf{k})$.

In column format:

$$\mathbf{r} = \begin{pmatrix} 4 \\ -5 \\ 1 \end{pmatrix} + t \begin{pmatrix} 3 \\ 4 \\ -2 \end{pmatrix}.$$

> **KEY POINT**
>
> A vector equation of the line through two fixed points A and B, with position vectors **a** and **b** is given by $\mathbf{r} = \mathbf{a} + t(\mathbf{b} - \mathbf{a})$ where t is a scalar parameter.
>
>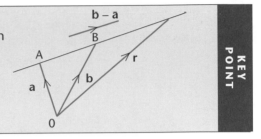

Example
Find a vector equation of the line through the points $P(3, 1, -4)$ and $Q(-2, 5, 1)$.

Direction of line PQ:

$$\overrightarrow{PQ} = \mathbf{q} - \mathbf{p} = -2\mathbf{i} + 5\mathbf{j} + \mathbf{k} - (3\mathbf{i} + \mathbf{j} - 4\mathbf{k}) = -5\mathbf{i} + 4\mathbf{j} + 5\mathbf{k}$$

A vector equation of PQ is $\mathbf{r} = 3\mathbf{i} + \mathbf{j} - 4\mathbf{k} + t(-5\mathbf{i} + 4\mathbf{j} + 5\mathbf{k})$.

Angles between two lines

To find the angle between two lines, find the angle between their direction vectors.

Pairs of lines

Core 4

In 2-dimensions, a pair of lines are either parallel or they intersect.
In 3-dimensions, a pair of lines are parallel, or they intersect or they are **skew**.

If two lines are parallel, then their direction vectors are multiples of each other, as in this **example**:
$\mathbf{r} = 3\mathbf{i} - 5\mathbf{j} + 2\mathbf{k} + \lambda(\mathbf{i} + 2\mathbf{j} - \mathbf{k})$ and $\mathbf{r} = 4\mathbf{i} + \mathbf{j} - 2\mathbf{k} + \mu(3\mathbf{i} + 6\mathbf{j} - 3\mathbf{k})$.

The two lines $\mathbf{r} = \mathbf{a} + \lambda\mathbf{b}$ and $\mathbf{r} = \mathbf{c} + \mu\mathbf{d}$ intersect if unique values of λ and μ can be found such that $\mathbf{a} + \lambda\mathbf{b} = \mathbf{c} + \mu\mathbf{d}$.
If unique values cannot be found, then the two lines are **skew**.

Example
Investigate whether the lines $\mathbf{r}_1 = 2\mathbf{i} + \mathbf{j} - 3\mathbf{k} + \lambda(4\mathbf{i} + 6\mathbf{j} - \mathbf{k})$ and $\mathbf{r}_2 = 3\mathbf{i} - 2\mathbf{k} + \mu(4\mathbf{i} + \mathbf{j} + 3\mathbf{k})$ intersect or whether they are skew.

If the lines intersect, then $\mathbf{r}_1 = \mathbf{r}_2$ will have a unique solution for λ and μ.

$\mathbf{r}_1 = \mathbf{r}_2 \implies 2\mathbf{i} + \mathbf{j} - 3\mathbf{k} + \lambda(4\mathbf{i} + 6\mathbf{j} - \mathbf{k}) = 3\mathbf{i} - 2\mathbf{k} + \mu(4\mathbf{i} + \mathbf{j} + 3\mathbf{k})$

i.e. $(2 + 4\lambda)\mathbf{i} + (1 + 6\lambda)\mathbf{j} + (-3 - \lambda)\mathbf{k} = (3 + 4\mu)\mathbf{i} + (0 + \mu)\mathbf{j} + (-2 + 3\mu)\mathbf{k}$

Note that the lines are not parallel, since their direction vectors are not multiples of each other.

Equating coefficients:

$$\begin{aligned} 2 + 4\lambda &= 3 + 4\mu \quad (1) \\ 1 + 6\lambda &= \mu \quad (2) \\ -3 - \lambda &= -2 + 3\mu \quad (3) \end{aligned}$$

Solving (1) and (2) simultaneously gives:

$\lambda = -0.25$ and $\mu = -0.5$
Check whether these satisfy (3):
LHS $= -3 - (-0.25) = -2.75$
RHS $= -2 + 3(-0.5) = -3.5$

$\therefore \lambda = -0.25$ and $\mu = -0.5$ do not satisfy all three equations.
Since there is not a unique solution for λ and μ, the lines do not intersect. They are skew.

Progress check

Core 4

1 Write down the position vector **r** of a point with coordinates (5, −2).

Core 4

2 Find the magnitude of the vector $\mathbf{r} = 5\mathbf{i} - 12\mathbf{j}$ and give its direction relative to **i**.

Core 4

3 Find the angle between the vectors $\mathbf{a} = 2\mathbf{i} + 3\mathbf{j} - \mathbf{k}$ and $\mathbf{b} = \mathbf{i} + \mathbf{j} + 2\mathbf{k}$.

Core 4

4 (a) Give a vector equation of the line which is parallel to $4\mathbf{i} - \mathbf{j} + 2\mathbf{k}$ and goes through the point with coordinates (3, 0, 1).
   (b) Give a vector equation of the line through (4, −1, 1) and (−3, 2, −2).

Core 4

5 (a) Find the point of intersection of the lines

   $$\mathbf{r} = 2\mathbf{i} - 3\mathbf{j} + \mu(\mathbf{i} - 2\mathbf{j}) \text{ and } \mathbf{r} = \mathbf{i} - 2\mathbf{j} + \lambda(2\mathbf{i} + 3\mathbf{j}).$$

   (b) Investigate whether the following lines intersect or whether they are skew. If they intersect, find their point of intersection.

   $$\mathbf{r} = \mathbf{i} + 2\mathbf{j} - 5\mathbf{k} + s(3\mathbf{i} + \mathbf{j} + \mathbf{k}) \text{ and } \mathbf{r} = 2\mathbf{i} - \mathbf{j} + 2\mathbf{k} + t(-\mathbf{i} - \mathbf{j} + \mathbf{k})$$

   (c) Find the angle between the lines given in (b).

5 (a) $(13\frac{2}{5}, -1\frac{3}{5})$ (b) Intersect at $(7, 4, -3)$ (c) $121.5°$
4 (a) $\mathbf{r} = 3\mathbf{i} + \mathbf{k} + t(4\mathbf{i} - \mathbf{j} + 2\mathbf{k})$ (b) $\mathbf{r} = 4\mathbf{i} - \mathbf{j} + \mathbf{k} + t(-7\mathbf{i} + 3\mathbf{j} - 3\mathbf{k})$
  (Answers are not unique, other answers are possible.)
3 $70.9°$
2 $|\mathbf{r}| = 13$, $67.4°$ clockwise.
1 $\mathbf{r} = 5\mathbf{i} - 2\mathbf{j}$

Sample questions and model answers

1

(a) Express $f(x) = \dfrac{5x+1}{(x+2)(x-1)^2}$ in partial fractions.

(b) Hence show that $\displaystyle\int_{-1}^{0} f(x)\,dx = 1 - 2\ln 2$.

> Remember the format when there is a repeated factor.

> Take care with the denominator.

(a) Let $\dfrac{5x+1}{(x+2)(x-1)^2} \equiv \dfrac{A}{x+2} + \dfrac{B}{x-1} + \dfrac{C}{(x-1)^2}$

$\Rightarrow \dfrac{5x+1}{(x+2)(x-1)^2} \equiv \dfrac{A(x-1)^2 + B(x+2)(x-1) + C(x+2)}{(x+2)(x-1)^2}$

$\Rightarrow \quad 5x+1 \equiv A(x-1)^2 + B(x+2)(x-1) + C(x+2)$

> Putting $x = 1$ makes the factor $(x-1)$ equal to zero.

> You could equate the x term or the constant term if you wish.

Let $x = 1$: $\quad 6 = 3C \Rightarrow C = 2$

Let $x = -2$: $\quad -9 = 9A \Rightarrow A = -1$

Equate x^2 terms: $\quad 0 = A + B \Rightarrow B = 1$

$\therefore \; f(x) = -\dfrac{1}{x+2} + \dfrac{1}{x-1} + \dfrac{2}{(x-1)^2}$

(b) $\displaystyle\int_{-1}^{0} f(x)\,dx = \int_{-1}^{0}\left(-\dfrac{1}{x+2} + \dfrac{1}{x-1} + \dfrac{2}{(x-1)^2}\right)dx$

> Remember the modulus sign.

$= \left[-\ln|x+2| + \ln|x-1| - 2(x-1)^{-1}\right]_{-1}^{0}$

$= -\ln 2 + \ln|-1| + 2 - (-\ln 1 + \ln|-2| + 1)$

$= -\ln 2 + 2 - \ln 2 - 1$

$= 1 - 2\ln 2$

> $\ln 1 = \ln|-1| = 0$
> $\ln|-2| = \ln 2$.

2

Expand $(1-2x)^{-3}$ in ascending powers of x as far as the term in x^3 and state the set of values of x for which the expansion is valid.

> Take care with the negatives.

> Remember to include the restriction on x, even when it is not requested specifically in the question.

$(1-2x)^{-3} = 1 + (-3)(-2x) + \dfrac{(-3)(-4)}{2!}(-2x)^2 + \dfrac{(-3)(-4)(-5)}{3!}(-2x)^3 + \ldots$

$= 1 + 6x + 24x^2 + 80x^3 + \ldots \quad$ provided $|2x| < 1$, i.e. $|x| < 0.5$

Sample questions and model answers (continued)

Core 4

3

Use the substitution $x = \tan u$ to show that $\displaystyle\int_0^1 \frac{1}{(1+x^2)^2}\,dx = \int_0^{\frac{\pi}{4}} \cos^2 u\,du.$

Hence show that $\displaystyle\int_0^1 \frac{1}{(1+x^2)^2}\,dx = \frac{\pi}{8} + \frac{1}{4}.$

Show this essential side working.

Remember to change the x limits to u limits.

$$\int_0^1 \frac{1}{(1+x^2)^2}\,dx = \int_{x=0}^{x=1} \frac{1}{(1+x^2)^2}\frac{dx}{du}\,du$$

$$= \int_0^{\frac{\pi}{4}} \frac{1}{(\sec^2 u)^2}\sec^2 u\,du$$

$$= \int_0^{\frac{\pi}{4}} \frac{1}{\sec^2 u}\,du$$

To integrate even powers of $\sin x$ or $\cos x$, change to double angles using $\cos 2x$ formula.

$$= \int_0^{\frac{\pi}{4}} \cos^2 u\,du$$

Let $x = \tan u$

$$\Rightarrow \frac{dx}{du} = \sec^2 u$$

$$1 + x^2 = 1 + \tan^2 u$$
$$= \sec^2 u$$

Limits:

When $x = 0$, $\tan u = 0 \Rightarrow u = 0$

When $x = 1$, $\tan u = 1 \Rightarrow u = \dfrac{\pi}{4}$

$$\int_0^{\frac{\pi}{4}} \cos^2 u\,du = \frac{1}{2}\int_0^{\frac{\pi}{4}} (1 + \cos 2u)\,du$$

$$= \frac{1}{2}\left[u + \frac{1}{2}\sin 2u\right]_0^{\frac{\pi}{4}}$$

$$= \frac{1}{2}\left(\frac{\pi}{4} + \frac{1}{2} - 0\right)$$

$$= \frac{\pi}{8} + \frac{1}{4}$$

$\cos 2u = 2\cos^2 u - 1$

$\Rightarrow \cos^2 u = \frac{1}{2}(1 + \cos 2u)$

Core 3

4

(a) Solve the inequality $|2x - 3| < 11$

(b) Sketch the graph of $y = |f(x)|$ where $f(x) = (x-1)(x-2)(x-3)$

(a) $|2x - 3| < 11$

This first step is important.

$$\Rightarrow -11 < 2x - 3 < 11$$

You can simplify the inequality by doing the same thing to each part.

$$\Rightarrow -8 < 2x < 14$$

Add 3 to each part.

$$\Rightarrow -4 < x < 7.$$

Divide each part by 2.

(b)

Parts of the graph of $y = f(x)$ that would be below the x-axis are reflected in the x-axis to give the graph of $y = |f(x)|$.

Sample questions and model answers (continued)

Core 3

5

The functions f and g are defined by:

$$f(x) = \sqrt{x-1} \quad x \geqslant 1$$
$$g(x) = 2x+3 \quad x \in \mathbb{R}$$

(a) Find an expression for $f^{-1}(x)$ and state its domain and range.

(b) Find the value of:

   (i) $fg(7)$

   (ii) $gf(10)$

   (iii) $f^{-1}g^{-1}(15)$

> Set $y = f(x)$ and rearrange to make x the subject.

(a) $y = \sqrt{x-1}$

$\Rightarrow y^2 = x - 1$

$\Rightarrow x = y^2 + 1$

$\Rightarrow f^{-1}(x) = x^2 + 1$

> The domain of f^{-1} is given by the range of f.

The domain of f^{-1} is $x \geqslant 0$ and the range of f^{-1} is $x \geqslant 1$.

> The range of f^{-1} is given by the domain of f.

> The order in which you apply the functions is important.

(b) (i) $fg(7) = f(17) = 4$

   (ii) $gf(10) = g(3) = 9$

   (iii) $f^{-1}g^{-1}(15) = f^{-1}(6) = 37$

Core 4

6

The equation of a straight line l is

$$\mathbf{r} = 4\mathbf{i} - 2\mathbf{k} + t(3\mathbf{i} + \mathbf{k}) \qquad \text{where } t \text{ is a scalar parameter.}$$

Points A, B and C lie on the line l. Referred to the origin O, they have position vectors $\mathbf{a}$, $\mathbf{b}$ and $\mathbf{c}$ respectively.

(a)   Find $\mathbf{a}$ and $\mathbf{b}$ and angle AOB, given that distance $OA = OB = 10$ units.

(b)   Find $\mathbf{c}$, given that OC is perpendicular to l.

(a) $\underline{a} = (4 + 3t)\underline{i} + (-2 + t)\underline{k}$ for some value t.   (1)

> Distance $OA = |\mathbf{a}|$.

Since $|\underline{a}| = 10$,

$$\sqrt{(4+3t)^2 + (-2+t)^2} = 10$$
$$16 + 24t + 9t^2 + 4 - 4t + t^2 = 100$$
$$10t^2 + 20t - 80 = 0$$
$$t^2 + 2t - 8 = 0$$
$$(t-2)(t+4) = 0$$

$\Rightarrow t = 2$ or $t = -4$

Sample questions and model answers (continued)

This gives the position vectors of **a** and **b**.

Substituting for t in (1) gives

$\underline{a} = 10\underline{i}$ and $\underline{b} = -8\underline{i} - 6\underline{k}$

Let angle AOB be θ.

$|\underline{a}| = |\underline{b}| = 10$ and $\underline{a} \cdot \underline{b} = (10\underline{i}) \cdot (-8\underline{i} - 6\underline{k}) = -80$

$\underline{a} \cdot \underline{b} = |\underline{a}||\underline{b}|\cos\theta$

$\implies \cos\theta = \dfrac{-80}{100} = -0.8$

$\implies \theta = 143.1°$ (1 d.p.)

(b) $\underline{c} = (4 + 3t)\underline{i} + (-2 + t)\underline{k}$ for some value t. (2)

Since $\underline{c}$ is perpendicular to l,

$((4 + 3t)\underline{i} + (-2 + t)\underline{k}) \cdot (3\underline{i} + \underline{k}) = 0$

$\implies 12 + 9t - 2 + t = 0$

$\implies t = -1$

Substituting for t in (2) gives $\underline{c} = \underline{i} - 3\underline{k}$.

3i + **k** is the direction vector of l and the scalar product of perpendicular vectors is zero.

Practice examination questions

Core 4

1 It is given that $f(x) = \dfrac{1}{(1+x)^2} + \sqrt{4+x}$.

Show that if x^3 and higher powers are ignored, $f(x) \approx a + bx + cx^2$ and find the values of a, b and c.

State the values of x for which the expansion is valid.

Core 3

2 (a) Find $\dfrac{dy}{dx}$ when $y = \cos^2 x \sin x$, expressing your answer in terms of $\cos x$.

Part (b) Core 4

(b) Using the substitution $u = \cos x$, or otherwise, find $\displaystyle\int \cos^2 x \sin x \, dx$.

Core 3

3 A curve has equation $y = \dfrac{x^2 - 4}{x + 1}$.

The equation of the normal to the curve at $(2, 0)$ is $ax + by + c = 0$, where a, b and c are integers.

Find the values of a, b, and c.

Core 3

4 The functions f and g are defined by:

$f : x \mapsto 5x - 7 \qquad x \in \mathbb{R}$

$g : x \mapsto (x+1)(x-1) \quad x \in \mathbb{R}$.

(a) Find the range of g.

(b) Find an expression for $fg(x)$.

(c) Determine the values of x for which $fg(x) = f(x)$.

Core 3

5 (a) Express $15 \sin x + 8 \cos x$ in the form $R \sin(x + \alpha)$ where $0 < \alpha < 90°$.

(b) State the maximum value of $15 \sin x + 8 \cos x$.

(c) Solve the equation $15 \sin x + 8 \cos x = 12$.

Give all the solutions in the interval $0 < x < 360°$.

Core 4

6 Expressing y in terms of x, solve the differential equation

$$x \frac{dy}{dx} = y + yx,$$

given that $y = 4$ when $x = 2$.

Core 3

7 (a) Use the chain rule to differentiate:

(i) $y = (4x - 3)^9$

(ii) $y = \ln(5 - 2x)$

Part (b) Core 4

(b) Using your results from part (a):

(i) Write down an expression for $\displaystyle\int (4x - 3)^8 \, dx$

(ii) Calculate $\displaystyle\int_3^4 \frac{1}{5 - 2x} \, dx$

Practice examination questions (continued)

Core 4

8 (a) Use integration by parts to evaluate $\displaystyle\int_0^1 x\,e^{-2x}\,dx$.

(b) The diagram shows the graph of $y = x\,e^{-x}$.

The curve has a maximum point at A.

The shaded region R is the area bounded by the curve, the x-axis and the vertical line through A.

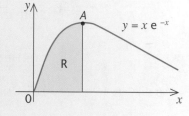

(i) Show that the coordinates of A are $(1, e^{-1})$.
(ii) The region R is rotated completely about the x-axis.
Calculate the volume of the solid generated.

Core 4

9 In a biology experiment, the growth of a population is being investigated.

A simple model is set up in which the rate of increase of the population at time t is proportional to the number, P, in the population at that time.

It is known that when $t = 0$, $P = 500$ and when $t = 10$, $P = 1000$.

Obtain the relationship between P and t and estimate the size of the population, according to the model, when $t = 20$.

Core 4

10 (a) Find the equation of the normal to the curve $x^3 + xy + y^2 = 7$ at the point $(1, 2)$.

(b) Find the coordinates of the two stationary points on the curve $x = t^2$, $y = t + \dfrac{1}{t}$.

Core 4

11 The region R, shaded in the diagram, is bounded by the curve $y = \sin 2x$ and the x-axis.

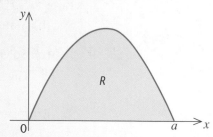

(a) Find the value of a.
(b) Show that the area of R is 1 square unit.
(c) Find the volume of the solid formed when R is rotated completely about the x-axis.

Core 4

12 (a) Find a vector equation of the line AB through $A(2, 2, 0)$ and $B(2, 3, 1)$.

(b) Show that the line AB and the line $\mathbf{r} = 2\mathbf{j} + \mathbf{k} + \lambda(\mathbf{i} + \mathbf{k})$, where λ is a scalar parameter, have no common point.

(c) Find the acute angle between the two lines.

Chapter 2

Statistics 2

The following topics are covered in this chapter:

- The binomial and Poisson distributions
- Continuous random variables
- Sampling
- Hypothesis tests

2.1 The binomial and Poisson distributions

After studying this section you should be able to:

- understand the binomial and Poisson distributions and select the appropriate model for a particular situation
- use the Poisson approximation to the binomial distribution
- use the normal approximation to the binomial distribution
- use the normal approximation to the Poisson distribution

LEARNING SUMMARY

The binomial distribution

Suppose that n **independent trials** are carried out, each with a **constant probability p of success.**

n and p are the parameters of the distribution.

If X is the number of successes in n trials, then X follows a **binomial distribution** and this is written $X \sim \text{B}(n, p)$.

Use nCr key on the calculator for $\binom{n}{x}$.

Probability	$P(X = x) = \binom{n}{x} p^x (1 - p)^{n-x}$	$x = 0, 1, 2, \ldots, n$
Mean	$\mu = np$	
Variance	$\sigma^2 = np(1 - p)$	
Standard deviation	$\sigma = \sqrt{np(1 - p)}$	

$\sigma = \sqrt{\text{variance}}$

KEY POINT

For some values of n and p you can use **cumulative Binomial probability tables** to find $P(X \leqslant x)$ and from this you can find other probabilities, for example:

$P(X < 3) = P(X \leqslant 2)$	0 1 2 3 ...
$P(X > 3) = 1 - P(X \leqslant 3)$	0 1 2 3 4 5 6 7 ...
$P(X \geqslant 3) = 1 - P(X \leqslant 2)$	0 1 2 3 4 5 ...
$P(X = 4) = P(X \leqslant 4) - P(X \leqslant 3)$	0 1 2 3 4 5 ...

Example
$X \sim \text{B}(10, 0.4)$.
Determine: (a) $P(X > 3)$ (b) $P(3 < X < 7)$ (c) the mean of X

You can use tables when $n = 10$, $p = 0.4$

(a) $P(X > 5) = 1 - P(X \leqslant 5) = 1 - 0.8338 = 0.1662$
(b) $P(3 < X < 7) = P(X \leqslant 6) - P(X \leqslant 3) = 0.9452 - 0.3823 = 0.5629$
(c) Mean $= np = 10 \times 0.4 = 4$

Example
$X \sim \text{B}(10, 0.44)$.

$p = 0.44$ is not in the binomial tables so you have to use the formula.

Determine: (a) $P(X = 7)$ (b) the standard deviation of X.

(a) $P(X = 7) = \binom{10}{7} 0.44^7 (1 - 0.44)^3 = 0.06728\ldots$

(b) Standard deviation $= \sigma = \sqrt{10 \times 0.44(1 - 0.44)} = 1.569\ldots$

The Poisson distribution

The **Poisson distribution** is used to model the number of occurrences, X, of an event in a given interval of space or time, when the events occur **independently** and **at random**, at a **constant average rate**.

λ is the parameter of the distribution.

If the mean number of occurrences in the interval is λ then $X \sim \text{Po}(\lambda)$.

> **KEY POINT**
>
> If $X \sim \text{Po}(\lambda)$:
>
> $$P(X = x) = e^{-\lambda}\frac{\lambda^x}{x!} \text{ for } x = 0, 1, 2, 3, \dots$$
>
> Mean, $\mu = \lambda$
>
> Variance $= \lambda$ and standard deviation $= \sqrt{\lambda}$

Mean = variance

This is similar to the using cumulative Binomial tables

For some values of λ you can use **cumulative Poisson probability** tables to find $P(X \leqslant x)$ and from this you can find other probabilities, for example:

$P(X < 2) = P(X \leqslant 1)$	0 1 2 3 …
$P(X > 4) = 1 - P(X \leqslant 4)$	0 1 2 3 4 5 6 7 …
$P(X \geqslant 3) = 1 - P(X \leqslant 2)$	0 1 2 3 4 5 …
$P(X = 4) = P(X \leqslant 4) - P(X \leqslant 3)$	0 1 2 3 4 5 …

Example

The number of letters received by the Adams family on a weekday may be modelled by a Poisson distribution with mean 3.75 and the number of letters received on a Saturday may be modelled by a Poisson distribution with mean 3. Determine the probability that:

(a) 4 letters are received on Monday

(b) more than 5 letters are received on Saturday.

$\lambda = 3.75$ is not in the Poisson tables so you have to use the formula.

(a) X is the number of letters received on Monday.

$$X \sim \text{Po}(3.75) \text{ so } P(X = 4) = = e^{-3.75}\frac{3.75^4}{4!} = 0.1937\dots = 0.194 \text{ (3 s.f.)}$$

$\lambda = 3$ is in the Poisson tables

(b) Y is the number of letters received on Saturday.

$$Y \sim \text{Po}(3) \text{ so } P(Y > 5) = 1 - P(Y \leqslant 5) = 1 - 0.9161 = 0.0839 \text{ (3 s.f.)}$$

The Poisson approximation to the binomial distribution

The binomial distribution $X \sim \text{B}(n, p)$ is used to model the number of successes in n independent trials when the probability of success, p, is constant.

The mean of the binomial distribution is np and the variance is $np(1 - p)$.

> **KEY POINT**
>
> When n is large ($n > 50$) and $np < 5$, $X \sim \text{B}(n, p)$ can be **approximated** by a Poisson distribution with the same mean, where $X \sim \text{Po}(np)$ approximately.

With these conditions, $np \approx np(1 - p)$. This fits in with the Poisson distribution where the mean and the variance are equal.

Example

The proportion of defective items produced by a machine is 2.5%. Find the probability of obtaining fewer than 3 defective items in a random sample of 100 items.

If X is the number of defective items in 100, then $X \sim B(100, 0.025)$.
The mean $np = 100 \times 0.025 = 2.5$
Since $n > 50$ and $np < 5$, use the Poisson approximation, $X \sim Po(2.5)$.

$P(X < 3) = P(X = 0) + P(X = 1) + P(X = 2)$

$$= e^{-2.5} + 2.5e^{-2.5} + \frac{2.5^2}{2!}e^{-2.5} = e^{-2.5}\left(1 + 2.5 + \frac{2.5^2}{2!}\right) = 0.54 \text{ (2 s.f.)}$$

$$P(X = x) = \frac{\lambda^x}{x!}e^{-\lambda}$$

It is often quicker to use cumulative probability tables, so make sure that you know how to use them.

If you have access to cumulative Poisson tables giving $P(X \leqslant x)$ for various values of x, then these tables can be used with $\lambda = 2.5$ and
$P(X < 3) = P(X \leqslant 2) = 0.5438 = 0.54 \text{ (2 s.f.)}$.

The normal approximation to the binomial distribution

> When n is large enough to ensure that $np > 5$ and $n(1 - p) > 5$, $X \sim B(n, p)$ can be **approximated** by a normal distribution with the same mean and the same variance, where $X \sim B(np, np(1 - p))$ approximately.
>
> **KEY POINT**

Key points from AS

- **The normal distribution**
  Revise AS pages 87–88

The larger the value of n and the closer that p is to 0.5, the better the approximation.

The normal distribution is continuous, whereas the binomial is discrete, so a **continuity correction** must be used.

Example

The probability that a person supports a particular organisation is 0.3. In a random sample of 50 people, find the probability that at least 20 support the organisation.

Always define the variable, giving its distribution if known.

If X is the number of people in 50 who support the organisation, then $X \sim B(50, 0.3)$.

Mean $np = 50 \times 0.3 = 15$, variance $np(1 - p) = 50 \times 0.3 \times 0.7 = 10.5$

$X \sim N(\mu, \sigma^2)$ with $\mu = 15$ and $\sigma^2 = 10.5 \Rightarrow \sigma = \sqrt{10.5}$

Conditions for normal approximation: $np = 15 > 5$, $n(1 - p) = 50 \times 0.7 = 35 > 5$

$\therefore \quad X \sim N(15, 10.5)$ approximately.

Applying the continuity correction,
$P(X \geqslant 20)$ becomes $P(X > 19.5)$.

Think of $X = 20$ as going from 19.5 to 20.5. Since 'at least 20' includes $X = 20$, find $P(X > 19.5)$ in the normal distribution.

$$P(X > 19.5) = P\left(Z > \frac{19.5 - 15}{\sqrt{10.5}}\right)$$

$$= P(Z > 1.39)$$

$$= 1 - \Phi(1.39) = 1 - 0.9177 = 0.082 \text{ (2 s.f.)}$$

Standardise
$Z = \frac{X - \mu}{\sigma}$
and then use standard normal tables.

Z: 0   1.39

The normal approximation to the Poisson distribution

The Poisson distribution, $X \sim Po(\lambda)$, is used to model the number of occurrences of an event, when events occur independently and at random, at a constant average rate. The mean is λ and the variance is λ.

> When λ is large ($\lambda > 15$ say), $X \sim Po(\lambda)$ can be approximated by a normal distribution with the same mean and variance, where $X \sim N(\lambda, \lambda)$ approximately.
>
> **KEY POINT**

Since the Poisson distribution is discrete, a continuity correction must be used.

Example

If $X \sim \text{Po}(20)$, find $P(15 < X < 22)$.

Since n is large, $X \sim N(20, 20)$ approximately.

$P(15 < X < 22)$ becomes $P(15.5 < X < 21.5)$.

You do not want to include 15 or 22 so go from 15.5 to 21.5

Apply the continuity correction.

Standardise the variables and then use normal tables.

$$P(15.5 < X < 21.5) = P\left(\frac{15.5 - 20}{\sqrt{20}} < Z < \frac{21.5 - 20}{\sqrt{20}}\right)$$

$$= P(-1.01 < Z < 0.34)$$

$$= \Phi(0.34) - \Phi(-1.01)$$

$$= \Phi(0.34) - (1 - \Phi(1.01))$$

$$= 0.6331 - (1 - 0.8438)$$

$$= 0.477 \text{ (3 s.f.)}$$

Z: -1.01 0 0.34

Progress check

1 Given that $X \sim B(20, 0.3)$ write down (a) $E(X)$ (b) $\text{Var}(X)$.

2 The variable X follows a binomial distribution where $X \sim B(n, p)$.
   (a) If $n = 12$ and $p = 0.25$, use cumulative binomial tables to determine
       (i) $P(X < 5)$ (ii) $P(X \geqslant 8)$ (iii) $P(2 \leqslant X \leqslant 7)$
   (b) If $n = 16$ and the mean is 12.8, determine, by using appropriate formulae
       (i) p (ii) $P(X = 11)$

3 The variable X follows a Poisson distribution with standard deviation 2.
   (a) Write down the value of λ.
   (b) Use cumulative probability tables to find:
       (i) $P(X < 5)$ (ii) $P(X \geqslant 1)$ (iii) $P(X = 6)$.

4 The number of telephone calls to a switchboard may be modelled by a Poisson distribution with mean 3.9 calls per 5-minute interval. Determine, to 3 decimal places, the probability that there will be:
   (a) at least one call in a 5-minute interval
   (b) exactly 9 calls in a 10-minute interval.

5 $X \sim B(80, 0.06)$. Find, to 3 decimal places, $P(X = 4)$
   (a) by using the binomial distribution,
   (b) by using a suitable approximation.

6 A newspaper reports that 62% of adults have an e-mail address. A random sample of 40 adults was selected. Using a suitable approximation, find the probability that at least 30 had an e-mail address.

7 Telephone calls reach a switchboard independently and at random at a rate of 30 per hour. Find the probability that in a randomly selected period of one hour there are fewer than 20 calls.

5 (a) 0.186 (b) 0.182 6 0.063 (2 s.f.) 7 0.028 (2 s.f.)
3 (a) 4 (b) (i) 0.6288 (ii) 0.9817 (iii) 0.1042 4 (a) 0.980 (b) 0.121
1 (a) 6 (b) 4.2 2 (a) (i) 0.8424 (ii) 0.0028 (iii) 0.8388 (b) (i) 0.8 (ii) 0.00739

2.2 Continuous random variables

After studying this section you should be able to:

- *find probabilities and the mode, mean and variance of a continuous variable with probability density function $f(x)$*
- *find the cumulative function $F(x)$ and the median and quartiles*
- *find the probability density function from the cumulative distribution function*
- *understand the continuous uniform distribution*

LEARNING SUMMARY

The probability density function, $f(x)$

The continuous random variable X is defined by its **probability density function** $f(x)$, together with the values for which it is valid.

$P(X = c) = 0.$

$$P(c \leqslant X \leqslant d) = \int_c^d f(x)\,dx$$

The total probability is 1.

If $f(x)$ is valid for $a \leqslant x \leqslant b$, then $\int_a^b f(x)\,dx = 1$.

It is useful to sketch the curve when finding the mode.

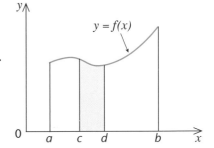

The **mode** is the x-value for which y is greatest in the defined interval.

The functions used most often are X and X^2 as E(X) is the mean and E(X^2) is used to find the variance.

The **expectation** of $g(x)$ is $E(g(x)) = \int_{\text{all } x} g(x)f(x)\,dx$, for any function $g(x)$.

The mean is the **expectation** of X, written E(X). It is often denoted by μ.

If $f(x)$ is given in stages, then the integration is carried out in stages also.

KEY POINT

Mean: $\quad E(X) = \mu = \displaystyle\int_{\text{all } x} x f(x)\,dx$

Variance: $\quad \text{Var}(X) = \sigma^2 = E(X^2) - \mu^2$, where $E(X^2) = \displaystyle\int_{\text{all } x} x^2 f(x)\,dx$

Example

The continuous random variable X has probability density function $f(x)$ given by $f(x) = \frac{1}{4}x$ for $1 \leqslant x \leqslant 3$. Find the mean μ, E(X^2) and the standard deviation σ.

$$\mu = E(X) = \int_{\text{all } x} x f(x)\,dx = \int_1^3 \tfrac{1}{4}x^2\,dx = \left[\frac{x^3}{12}\right]_1^3 = 2\tfrac{1}{6}$$

$$E(X^2) = \int_{\text{all } x} x^2 f(x)\,dx = \int_1^3 \tfrac{1}{4}x^3\,dx = \left[\frac{x^4}{16}\right]_1^3 = 5$$

$\sigma = \sqrt{\text{variance}}.$

$$\text{Var}(X) = 5 - (2\tfrac{1}{6})^2 = \tfrac{11}{36} \implies \sigma = \sqrt{\tfrac{11}{36}} = 0.553 \text{ (3 d.p.)}$$

69

The cumulative distribution function, $F(x)$

The **cumulative distribution function**, $F(x)$, gives the probability that X is less than a particular value, i.e. $P(X < x)$. It is found from the area under the probability curve:

KEY POINT

$F(x_1) = P(X < x_1)$.

$$F(x_1) = \int_a^{x_1} f(x)\,dx \qquad \text{for } a \leqslant x_1 \leqslant b$$

Sometimes the median is denoted by q_2.

The **median**, m, is the value 50% of the way through the distribution. It divides the total area in half, so $F(m) = 0.5$

The **lower quartile** q_1 is the value 25% of the way, so $F(q_1) = 0.25$
The **upper quartile** q_3 is the value 75% of the way, so $F(q_3) = 0.75$

The **interquartile range** = upper quartile – lower quartile.

Example

$f(x) = \frac{1}{4}x$ for $1 \leqslant x \leqslant 3$. Find the median, m, and the interquartile range.

x_1 is used here to denote any value of x between 1 and 3. Sometimes a different letter is used, such as t.

For $1 \leqslant x_1 \leqslant 3$, $F(x_1) = \int_1^x \frac{1}{4}x\,dx = \left[\dfrac{x^2}{8}\right]_1^{x_1} = \frac{1}{8}(x_1^2 - 1)$

$F(m) = 0.5 \Rightarrow \frac{1}{8}(m^2 - 1) = 0.5 \Rightarrow m^2 = 5,\ m = \sqrt{5}\ (= 2.236\ldots)$
$F(q_1) = 0.25 \Rightarrow \frac{1}{8}(q_1^2 - 1) = 0.25 \Rightarrow q_1^2 = 3,\ q_1 = \sqrt{3}\ (= 1.732\ldots)$
$F(q_3) = 0.75 \Rightarrow \frac{1}{8}(q_3^2 - 1) = 0.75 \Rightarrow q_3^2 = 7,\ q_3 = \sqrt{7}\ (= 2.645\ldots)$
Interquartile range $= q_3 - q_1 = \sqrt{7} - \sqrt{3} = 0.91$ (2 d.p.)

Differentiating the cumulative function gives the probability function.

The probability density function can be obtained from the cumulative distribution function:

KEY POINT

$$f(x) = \frac{d}{dx}F(x).$$

Example

Find $f(x)$ where $F(x) = \begin{cases} 0 & x \leqslant 0 \\ \frac{1}{8}x^3 & 0 \leqslant x \leqslant 2 \\ 1 & x \geqslant 2 \end{cases}$

$f(x) = \dfrac{d}{dx}(\frac{1}{8}x^3) = \frac{3}{8}x^2$ for $0 \leqslant x \leqslant 2$ and $f(x) = 0$ otherwise.

The uniform distribution

This is also known as the rectangular distribution.

If $f(x)$ is distributed **uniformly** in the interval $a \leqslant x \leqslant b$, then $f(x) = \dfrac{1}{b-a}$.

Mean $= E(X) = \dfrac{1}{2}(a+b)$

Variance $= Var(X) = \dfrac{1}{12}(b-a)^2$

$F(x_o) = \dfrac{x_o - a}{b-a}$

Progress check

1 $f(x) = 1 - 0.5x$, $0 \leqslant x \leqslant 2$.

   (a) Find $E(X)$, $E(X^2)$ and $Var(X)$.
   (b) Find the cumulative function $F(x)$, the median and the interquartile range.

2 $f(x)$ is distributed uniformly in the interval $1 \leqslant x \leqslant 6$. Find:

   (a) $P(1.8 < X < 3.2)$ (b) $E(X)$ (c) the standard deviation of X (d) $F(3.2)$

3 A continuous random variable X has cumulative distribution function $F(x)$ as shown. Find the probability density function $f(x)$.

$$F(x) = \begin{cases} 0 & x < 2 \\ \dfrac{1}{20}x^2 + \dfrac{1}{5}x - \dfrac{3}{5} & 2 \leqslant x \leqslant 4 \\ 1 & x > 4 \end{cases}$$

3 $f(x) = \frac{1}{10}x + \frac{1}{5}$ for $2 \leqslant x \leqslant 4$ and $f(x) = 0$ otherwise

2 (a) 0.28 (b) 3.5 (c) 1.44 (2 d.p.) (d) 0.44

1 (a) $\frac{2}{3}, \frac{2}{3}, \frac{2}{9}$ (b) $F(x) = 0$, $x < 0$, $F(x) = x - 0.25x^2$, $0 \leqslant x < 2$, $F(x) = 1$, $x > 2$; median = 0.586 (3 d.p.), IQR = 0.732 (3 d.p.)

2.3 Sampling

After studying this section you should be able to:

* understand the difference between a population and a sample
* understand the difference between a census and a sample survey
* understand the concept of a statistic and its sampling distribution

LEARNING
SUMMARY

Populations and samples

* A **population** is the entire collection of individuals or items under consideration.
* A **sample** is a selection of individuals or items from the population.
* These individual members are known as **sampling units**.
* A list of the sampling units is a **sampling frame**.

When you want information about a **population**, you could carry out a **census** of every member. The advantage is that you would have accurate and complete information. Disadvantages include cost and time requirements. Taking a census could also destroy the population. For example, if you were investigating the length of life of a calculator battery, carrying out a census would destroy the population.

More usually, a **sample** is taken. The sample should be representative of the whole population, so **bias** in the choice of sample members must be avoided.

In a **random sample** of size n, each member of the population has an equal chance of being selected. Also all subsets of the population of size n have an equal chance of constituting the sample.

Assigning a number to each member of a population and then drawing the numbers out of a hat is one method of obtaining a **simple random sample**. Another is to use **random number tables**, in which each digit has an equal chance of occurring.

If there is a periodic fault, then systematic sampling may fail to register it or may give it undue weight.

Random sampling from a very large population can be very laborious. It may be more convenient to carry out **systematic sampling**. This involves selecting every kth member of the population, for example, every 10th item from a particular machine on a production line.

The stratified sample can be constructed in proportion to the number of members in each stratum.

The method of **stratified sampling** is often used when the population is split into distinguishable layers or strata, such as students in each faculty in a college.

Statistics and sampling distributions

A **parameter** is a numerical property of a population such as the mean μ, the standard deviation σ or the proportion of successes p.

A **statistic** is a numerical property of a sample, such as the sample mean, $\overline{X}$.

In general, if $X_1, X_2, X_3, \ldots, X_n$ is a random sample size n from a population, then a statistic is a random variable consisting of any function of X that involves no other quantities.

Here are examples of statistics:

$$\overline{X} = \frac{\sum_{i=1}^{n} X_i}{n} \qquad S^2 = \frac{\Sigma(X_i - \overline{X})^2}{n-1} \qquad T = \frac{X_1 + X_2}{2}$$

The following are not statistics, as they contain parameters:

$$\frac{\sum_{i=1}^{n}(X_i - \mu)^2}{n} \qquad \sum_{i=1}^{n} \frac{X_i - u}{\sigma}$$

The distribution of the X_i, which is the same as the distribution of the population, determines the distribution of the statistic, called the **sampling distribution of the statistic**.

Progress check

1 A health centre wanted to assess patients' satisfaction with the services provided. Questionnaires were given to all patients who attended the health centre on a particular day.
   (a) State one advantage and one disadvantage of using a sample survey rather than a census.
   (b) Suggest a suitable sampling frame.
   (c) Identify the sampling units.

2 A random sample $X_1, X_2, \ldots X_n$ is taken from a population with unknown mean μ and unknown variance σ^2. A statistic is based on this sample.

State which, if any, of the following are **not** statistics:
$$T = \sum_{i=1}^{n} \frac{X_i^2}{\sigma} \qquad U = \text{Minimum of } \{X_1, X_2, \ldots X_n\} \qquad V = \frac{X_1 + X_2 + \ldots + X_n}{n}$$

1 (a) A census is quicker to carry out, but may not be representative.
(b) A list of all the names of all the patients who attended on that day.
(c) The individual patients.
2 T

2.4 Hypothesis tests

After studying this section you should be able to:

- *understand the language used in hypothesis testing*
- *perform tests for a binomial proportion*
- *perform a discrete test for a Poisson mean*

LEARNING SUMMARY

The language of hypothesis testing

When you want to know something about a population, you might perform a **hypothesis** or **significance test**.

A **null hypothesis**, H_0 is made about the population.
Then the **alternative hypothesis H_1**, the situation when H_0 is not true, is stated.

Example

When investigating the proportion p in a binomial distribution, you might make the null hypothesis that the proportion is 0.2. This is written H_0: p = 0.2

The alternative hypothesis would be one of the following:

Use p > 0.2 if you suspect an *increase*.

Use p ≠ 0.2 if you suspect a *change*.

- H_1: p > 0.2 (one-tailed, upper tail test)
- H_1: p < 0.2 (one-tailed, lower tail test)
- H_1: p ≠ 0.2 (two-tailed test).

Use p < 0.2 if you suspect a *decrease*.

The formula for the test statistic depends on the sampling distribution. Examples are shown in the following text.

A **test statistic** is defined and its distribution when H_0 is true is stated. Its value, based on information from a random sample taken from the population, is found. This is the **test value**.

The hypothesis test involves deciding whether or not the test value could have come from the distribution defined by the null hypothesis.
If the test value is in the main bulk of the distribution (the **acceptance region**) it is *likely* to have come from the distribution. If it is in the 'tail' end (the **rejection** or **critical region**), it is *unlikely* to have come from the distribution.

Probability theory is used to decide the placing of the boundary between 'likely' and 'unlikely'.

The decision rule (**rejection criterion**) is based on the **significance level** of the test, which is used to fix the boundaries for the rejection region. The boundaries are called **critical values**.

The significance level is a given as a percentage.

The levels used most often are 10%, 5% and 1%.

For **example**, for a significance level of 5%, the critical values are such that 5% of the distribution is in the critical (rejection) region,

i.e. $P(X >$ critical value$) = 0.05$, for a one-tailed, upper tail test
$P(X <$ critical value$) = 0.05$, for a one-tailed, lower tail test
$P(X >$ upper critical value$) = 0.025$, for a two-tailed test
$P(X <$ lower critical value$) = 0.025$, for a two-tailed test.

There are two critical values in a two-tailed test.

Depending on the position of the test value, the **decision** is made.
If the test value lies in the critical region, H_0 is rejected in favour of H_1.
If the test value is in the acceptance region, H_0 is not rejected.

A test statistic is said to be significant if it lies in the critical region. If it lies in the acceptance region, then it is not significant.

The **conclusion** is then stated in relation to the situation being tested.

Tests for a binomial proportion

This binomial test is used to test the proportion of successes, p, in a binomial population.

The test statistic is X, the number of successes, where $X \sim B(n, p)$. The critical values can be difficult to find, so it is easier to use a probability method to set up the decision rules, as follows:

For **example**, at the 5% level, the decision rules are:

- for a one-tailed (upper tail) test, reject H_0 if P($X \geqslant$ test value) < 0.05
- for a one-tailed (lower tail) test, reject H_0 if P($X \leqslant$ test value) < 0.05
- for a two-tailed test, reject H_0 if P($X \leqslant$ test value) < 0.025 or P($X \geqslant$ test value) < 0.025

Example

When a die was thrown 10 times, a six occurred 4 times. Is this evidence, at the 10% level of significance, that the die is biased in favour of sixes?

Let the probability of obtaining a six be p.
If X is the number of sixes in 10 throws, then $X \sim B(10, p)$.
H_0: $p = \frac{1}{6}$ (the die is fair), H_1: $p > \frac{1}{6}$ (the die is biased in favour of sixes).

> This is a one-tailed (upper tail) test.

If the null hypothesis is true, then $X \sim B(10, \frac{1}{6})$.
Reject H_0 if the test value of $x = 4$ lies in the critical region (the upper tail 10%)
i.e. reject H_0 if P($X \geqslant 4$) < 0.1
P($X \geqslant 4$) = 1 − P($X < 4$)

> Sometimes it may be possible to calculate the probability using cumulative binomial tables.

$$= 1 - \left(\left(\tfrac{5}{6}\right)^{10} + {}^{10}C_1\left(\tfrac{5}{6}\right)^9\left(\tfrac{1}{6}\right) + {}^{10}C_2\left(\tfrac{5}{6}\right)^8\left(\tfrac{1}{6}\right)^2 + {}^{10}C_3\left(\tfrac{5}{6}\right)^7\left(\tfrac{1}{6}\right)^3\right)$$
$$= 1 - 0.930 \ldots = 0.070 \text{ (2 s.f.)}.$$

Since P($X \geqslant 4$) < 0.1, H_0 is rejected.

There is evidence at the 10% level, that the die is biased in favour of sixes.

Test for a Poisson mean

The test statistic is X, the number of occurrences, where $X \sim Po(\lambda)$.
The method is similar to that used in the binomial test and the decision rules are the same.

Example

The office manager claims that the average number of telephone calls per minute received by her office switchboard is 4.5. Her supervisor maintains that it is less than 4.5. In a randomly selected minute, 2 calls were received. A hypothesis test is conducted at the 5% level. Whose claim is upheld?

Let X be the number of calls per minute. Assuming that calls occur randomly and independently, $X \sim Po(\lambda)$.

> This is a one-tailed (lower tail) test.

H_0: $\lambda = 4.5$, H_1: $\lambda < 4.5$. If the null hypothesis is true, $X \sim Po(4.5)$.

The test value is $x = 2$, so, at the 5% level, reject H_0 if P($X \leqslant 2$) < 0.05

> This could have been found using cumulative Poisson tables.

$$P(X \leqslant 2) = e^{-4.5} + 4.5e^{-4.5} + \frac{4.5^2}{2!}e^{-4.5} = e^{-4.5}\left(1 + 4.5 + \frac{4.5^2}{2!}\right) = 0.173 \ldots$$

Since P($X \leqslant 2$) > 0.05, the test value does not lie in the critical region and H_0 is not rejected. The office manager's claim that the mean is 4.5 is upheld.

Progress check

1 A random observation, x is taken from $X \sim \mathrm{B}(n, p)$.
   Test, at the 5% level, the null hypothesis that $p = 0.4$, against the alternative hypothesis that $p < 0.4$ if $n = 10$ and $x = 1$.

2 The number of breakdowns in a month of a machine is modelled by a Poisson distribution with mean 5.5. During the month following its annual service, the machine broke down twice.
   (a) Is there evidence, at the 10% level, of an improvement in the reliability of the machine?
   (b) Would your decision be the same if the test is carried out at the 5% level? Give a reason for your answer.

1 $P(X \leqslant 1) = 0.0464 < 0.05$, reject H_0, conclude $p < 0.4$.

2 (a) $P(X \leqslant 2) = 0.0884 < 0.1$. Evidence of improvement at 10% level.
   (b) No, since $0.0884 > 0.05$, there is no evidence of improvement at 5% level.

Sample questions and model answers

1

The continuous random variable X has probability density function $f(x)$ given by

$$f(x) = \begin{cases} kx(2-x) & 0 \leqslant x \leqslant 2 \\ 0 & \text{otherwise} \end{cases}$$

where k is a constant.

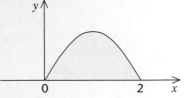

(a) Show that $k = \frac{3}{4}$ and find the exact value of $P(X > 1\frac{1}{2})$.

(b) Find $E(X)$ and $E(X^2)$.

(c) Show that the standard deviation, σ, is 0.447 correct to three decimal places.

The total area under the curve is 1.

As k is a constant, it can be taken outside the integration.

(a)
$$\int_0^2 f(x)\,dx = 1$$

$$\Rightarrow k\int_0^2 (2x - x^2)\,dx = 1$$

$$\Rightarrow k\left[x^2 - \frac{x^3}{3}\right]_0^2 = 1$$

$$\Rightarrow k(4 - \tfrac{8}{3} - 0) = 1$$

$$\Rightarrow k = \tfrac{3}{4}$$

$$f(x) = \tfrac{3}{4}(2x - x^2)$$

$$P(X > 1\tfrac{1}{2}) = \int_{1\frac{1}{2}}^2 f(x)\,dx$$

$$= \tfrac{3}{4}\int_{1\frac{1}{2}}^2 (2x - x^2)\,dx$$

$$= \tfrac{3}{4}\left[x^2 - \frac{x^3}{3}\right]_{1\frac{1}{2}}^2$$

Substitute the limits carefully, working in fractions to obtain the exact answer.

$$= \tfrac{3}{4}(4 - \tfrac{8}{3} - (\tfrac{9}{4} - \tfrac{9}{8}))$$

$$= \tfrac{5}{32}$$

If you do not spot that $f(x)$ is symmetrical about $x = 1$, find $E(X)$ by integration.

(b) By symmetry, $E(X) = 1$

$$E(X^2) = \int_0^2 x^2 f(x)\,dx$$

$$= \tfrac{3}{4}\int_0^2 (2x^3 - x^4)\,dx$$

$$= \tfrac{3}{4}\left[\frac{2x^4}{4} - \frac{x^5}{5}\right]_0^2$$

$$= \tfrac{3}{4}(8 - \tfrac{32}{5})$$

$$= 1.2$$

(c) $\text{Var}(X) = E(X^2) - [E(X)]^2$

$[E(X)]^2$ is often written $E^2(X)$.

$$= 1.2 - 1$$

$$= 0.2$$

$$\sigma = \sqrt{0.2} = 0.447 \text{ (3 d.p.)}$$

Practice examination questions

1 The probability that a person suffers from a particular health condition is 0.01

(a) Using a suitable approximation, find the probability that in a randomly chosen sample of 80 people, fewer than 3 suffer from the health condition.

(b) Find the minimum sample size in order that the probability of including at least one with the health condition is greater than 95%.

2 The continuous random variable X is distributed uniformly in the interval $2 \leqslant x \leqslant 12$.

(a) Find the probability density function, $f(x)$.

(b) Calculate the probability that X lies within one standard deviation of the mean, giving your answer correct to 2 decimal places.

3 A beetle infestation is discovered in the trees in a small copse. If more than 35% of the trees are infected it is impractical to treat the trees with chemicals and they will have to be felled.

The representative from the ministry checked a random sample of 10 trees and found that 6 were infected. Stating any assumptions, use a hypothesis test, at the 10% significance level, to decide whether the infestation should be treated with chemicals or the trees in the copse felled.

4 The continuous random variable X has probability function $f(x)$ where $f(x) = k(x - x^3)$, $0 \leqslant x \leqslant 1$ and $f(x) = 0$ otherwise.

(a) Show that $k = 4$.

(b) Find E(X).

(c) Find P($X < 0.5$).

(d) Find Var(X).

5 The continuous random variable X has cumulative distribution function given by

$$F(x) = \begin{cases} 0 & x \leqslant 0 \\ 2cx & 0 \leqslant x \leqslant 1 \\ c(4x - x^2 - 1) & 1 \leqslant x \leqslant 2 \\ 1 & x \geqslant 2 \end{cases}$$

(a) Show that $c = \frac{1}{3}$.

(b) Show that P($X > 1\frac{1}{2}$) = $\frac{1}{12}$.

(c) Find the probability function $f(x)$.

(d) Calculate E(X).

Chapter 3

Mechanics 2

The following topics are covered in this chapter:

- Kinematics
- Centre of mass
- Work, energy and power

- Collisions and restitution
- Statics of rigid bodies

3.1 Kinematics

After studying this section you should be able to:

- model the motion of a projectile moving under constant acceleration
- understand the limitations of the model
- solve problems by considering, separately, the vertical and horizontal components of velocity
- set up and solve a differential equation representing acceleration in a straight line, where the acceleration is given as a function of displacement or time
- solve problems which can be modelled as the linear motion of a particle under the action of a variable force
- differentiate and integrate a vector with respect to time to solve problems
- derive and apply the trajectory formula

Kinematics is the branch of mathematics that deals with the motion of a particle in terms of displacement, velocity and acceleration, without considering the forces that may be required to cause the motion.

The model and motion with constant acceleration

In the usual **particle** model for projectiles, air resistance is ignored and the only force taken to act on the particle is its weight which is constant.
According to this model:

> Check the value of g used on your exam paper. Make sure that you use the given value or you may lose marks.

- The horizontal component of velocity remains constant throughout the motion.
- The vertical component of velocity is subject to a constant downward acceleration of g ms^{-2}.

For motion in a straight line, the distinction between **distance** and **displacement** is only that displacement may be positive or negative to indicate a sense of direction, whereas distance is always taken to be positive.

The same distinction applies to **speed** and **velocity**. Velocity in a straight line may be positive or negative, depending on the direction of movement, but speed is always taken to be positive.

The accuracy of the model, in predicting the motion of a projectile, depends on the extent to which the initial assumptions are satisfied.

For example, the model will provide accurate information about a projectile such as a stone or a dart over a short distance, but will give poor results for light objects that are affected more by air resistance.

When using the model, a first step is usually to resolve the velocity of a projectile into horizontal and vertical components. The two parts are then treated separately.

A particle moving with speed v at angle θ to the horizontal has vertical component $v \sin \theta$ and horizontal component $v \cos \theta$.

If you know the horizontal and vertical components of a velocity then you can combine them to find its magnitude (speed) and direction.

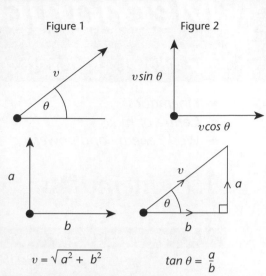

Figure 1   Figure 2

$v = \sqrt{a^2 + b^2}$ $\tan \theta = \dfrac{a}{b}$

Projectiles

In the standard model for projectiles, air resistance is ignored and the acceleration due to gravity g m s^{-2} is taken to be constant. According to this model:

> - The only force acting on a projectile, once in flight, is its weight.
> - The horizontal component of velocity is constant.
> - The vertical component of velocity is subject to a constant acceleration g m s^{-2}.

KEY POINT

The approach used for many projectile problems is to resolve the initial velocity into horizontal and vertical components and to treat these two parts separately.

Example
A stone is thrown horizontally with a speed of 10 m s^{-1} from the top of a cliff. The cliff is 98 m high and the stone lands in the sea d m from its base. Take $g = 9.8$ m s^{-2} and show that $d = 20\sqrt{5}$.

Vertically: $u = 0$
$\qquad\qquad a = 9.8$
$\qquad\qquad s = 98.$
Using $\qquad s = ut + \frac{1}{2}at^2$
gives $\qquad 98 = 4.9t^2$
$\qquad\qquad \Rightarrow t = \sqrt{20} = 2\sqrt{5}.$
Horizontally: $d = 10 \times 2\sqrt{5}$
$\qquad\qquad \Rightarrow d = 20\sqrt{5}.$

The initial velocity of the projectile may be at angle θ to the horizontal.

Example
A ball is struck with velocity 20 m s^{-1} at 40° to the horizontal from a point 1 m above the ground. Find the maximum height reached by the ball. Take $g = 9.8$ m s^{-2}.

Vertically: $u = 20 \sin 40°$
$\qquad\qquad a = -9.8$
At max ht: $v = 0$
Using $\qquad v = u + at$
$\qquad\qquad 0 = 20 \sin 40° - 9.8t$
$\qquad\qquad \Rightarrow t = 1.3118 \ldots.$
Using $\qquad s = ut + \frac{1}{2}at^2.$
At max ht $\qquad s = 20 \sin 40° \times 1.3118 - 4.9 \times 1.3118^2$
$\qquad\qquad = 8.432 \ldots.$

The maximum height is 9.43 m to 2 d.p.

Motion with variable acceleration

In this case, the constant acceleration formulae do not apply and must not be used. Using x for displacement, v for velocity and a for acceleration:

$$\xrightarrow{\text{differentiate}}$$
$$\begin{array}{ccc} x & v & a \end{array}$$
$$\xleftarrow{\text{integrate}}$$

$$v = \frac{dx}{dt} \qquad a = \frac{dv}{dt} = \frac{d^2x}{dt^2}$$

$$x = \int v\,dt \qquad v = \int a\,dt.$$

Example

A particle P moves along the x-axis such that its velocity at time t is given by $v = 5t - t^2$. When $t = 0$, $x = 15$. Find a formula for:

(a) the acceleration of the particle at time t.

(b) the position of the particle at time t.

(a) $v = 5t - t^2 \implies \dfrac{dv}{dt} = 5 - 2t.$

The acceleration of the particle at time t is given by $a = 5 - 2t$.

(b) $\qquad x = \int v\,dt = \int 5t - t^2\,dt$

$\implies \quad x = \dfrac{5t^2}{2} - \dfrac{t^3}{3} + c.$

When $t = 0$, $x = 15$

so $15 = 0 + c \implies c = 15.$

This gives the position of the particle at time t as $x = \dfrac{5t^2}{2} - \dfrac{t^3}{3} + 15.$

In the standard notation:

- displacement is represented by x

- velocity is represented by $v = \dot{x} = \dfrac{dx}{dt}.$

You may need to set up and solve equations of the form $\dfrac{dx}{dt} = f(x)$ or $\dfrac{dx}{dt} = f(t).$

- **Acceleration** is represented by $\ddot{x} = \dfrac{dv}{dt} = v\dfrac{dv}{dx}.$

> This is an important result and you may need to establish it in the exam.

You should recognise that by the chain rule for differentiation: $\dfrac{dv}{dt} = \dfrac{dv}{dx} \times \dfrac{dx}{dt} = v\dfrac{dv}{dx}.$

You may need to set up and solve equations of the form $\dfrac{dv}{dt} = f(t)$ or $v\dfrac{dv}{dx} = f(x).$

> This requires a differential equation because the acceleration is not constant.

Example

A particle P moves in the direction of the positive x-axis with acceleration $3\sqrt{x}$ ms^{-2} directed away from O, where OP $= x$ metres. When $x = 1$ the speed of P is 2 ms^{-1}. Find its speed when $x = 4$.

> See **separation of variables** page 47.

The acceleration is $v\dfrac{dv}{dx} = 3x^{\frac{1}{2}} \implies \int v\,dv = \int 3x^{\frac{1}{2}}\,dx$

This gives: $\qquad \dfrac{v^2}{2} = 2x^{\frac{3}{2}} + c$

When $x = 1$, $v = 2$ so $2 = 2 + c \Rightarrow c = 0$.

so: $$v^2 = 4x^{\frac{3}{2}}$$

When $x = 4$, $v^2 = 32 \Rightarrow v = 4\sqrt{2}$

The particle has speed $4\sqrt{2}$ ms^{-1} when $x = 4$.

Differentiation of vectors

A vector of the form $\mathbf{r} = f(t)\mathbf{i} + g(t)\mathbf{j}$ may be differentiated with respect to time to give $\dot{\mathbf{r}} = f'(t)\mathbf{i} + g'(t)\mathbf{j}$

For example, if $\mathbf{r} = 3t^2\mathbf{i} + \sqrt{t}\mathbf{j}$ then $\dot{\mathbf{r}} = 6t\mathbf{i} + \dfrac{1}{2\sqrt{t}}\mathbf{j}$.

> Differentiate the coefficients of **i** and **j** with respect to *t*.

If $\mathbf{r}$ represents the position vector of some moving point at time t then $\dot{\mathbf{r}}$ represents the corresponding velocity at time t.

$\dot{\mathbf{r}}$ may be differentiated to give $\ddot{\mathbf{r}}$ which represents the acceleration of the moving point at time t.

Example
A particle P has position vector $\mathbf{r}$ metres relative to a fixed point O, at time t seconds, where $\mathbf{r} = (2t^2 + 5t)\mathbf{i} + t^3\mathbf{j}$ and $t \geqslant 0$.
Find:
(a) the velocity of P at time t seconds
(b) the speed of P when $t = 3$
(c) the time when P is moving parallel to the vector $\mathbf{i} + 3\mathbf{j}$.

(a) The velocity of P in metres per second at time t is given by $\dot{\mathbf{r}} = (4t + 5)\mathbf{i} + 3t^2\mathbf{j}$.
(b) When $t = 3$, $\dot{\mathbf{r}} = 17\mathbf{i} + 27\mathbf{j}$.

> The speed of P is the magnitude of its velocity.

The speed of P when $t = 3$ is $\sqrt{17^2 + 27^2}$ ms^{-1} = 31.9 ms^{-1} to 3 s.f.

(c) When P is moving parallel to $\mathbf{i} + 3\mathbf{j}$ the component of velocity in the **j** direction must be 3 times the component in the **i** direction.

This gives:
$$3t^2 = 3(4t + 5)$$
so: $$t^2 = 4t + 5$$
$$t^2 - 4t - 5 = 0$$
$$(t - 5)(t + 1) = 0$$
$$t = 5 \text{ or } t = -1$$
However, $t \geqslant 0$ so the only solution is $t = 5$.

Integration of vectors

In general, $\displaystyle\int (f(t)\mathbf{i} + g(t)\mathbf{j})dt = \left(\int f(t)dt\right)\mathbf{i} + \left(\int g(t)dt\right)\mathbf{j}$

> Don't be put off by the complicated looking formula, just look at what it means.

So, to integrate a vector with respect to t, integrate the coefficients of **i** and **j** with respect to t. **Note that a constant of integration is needed and that this is itself a vector.**

Example

Find $\mathbf{v} = \displaystyle\int \mathbf{a}\, dt$ given that $\mathbf{a} = 2t\mathbf{i} - 6t^2\mathbf{j}$ and that $\mathbf{v} = 3\mathbf{i} + \mathbf{j}$ when $t = 1$.

$$\mathbf{v} = \int (2t\mathbf{i} - 6t^2\mathbf{j})dt = t^2\mathbf{i} - 2t^3\mathbf{j} + \mathbf{c}$$

When $t = 1$: $\mathbf{v} = \mathbf{i} - 2\mathbf{j} + \mathbf{c} = 3\mathbf{i} + \mathbf{j}$

so: $\mathbf{c} = 2\mathbf{i} + 3\mathbf{j}$

This gives: $\mathbf{v} = t^2\mathbf{i} - 2t^3\mathbf{j} + 2\mathbf{i} + 3\mathbf{j}$
or $\mathbf{v} = (t^2 + 2)\mathbf{i} + (3 - 2t^3)\mathbf{j}$

In the example, if $\mathbf{a}$ represents acceleration then $\mathbf{v}$ represents velocity.

> **KEY POINT**
>
> Vectors for position $\mathbf{r}$, velocity $\mathbf{v}$ and acceleration $\mathbf{a}$ are linked through differentiation and integration.
>
> $\mathbf{v} = \dot{\mathbf{r}}$ $\qquad \mathbf{v} = \int \mathbf{a}\, dt$ $\qquad$ differentiate
>
> $\qquad\qquad\qquad\qquad\qquad\qquad\qquad$ r $\quad$ v $\quad$ a
>
> $\mathbf{a} = \dot{\mathbf{v}} = \ddot{\mathbf{r}}$ $\qquad \mathbf{r} = \int \mathbf{v}\, dt$ $\qquad$ integrate

Example

> 'Initially' means when $t = 0$.

A particle P, initially at rest, has acceleration $\mathbf{a} = 2t\mathbf{i} + 3\sqrt{t}\mathbf{j}$ ms^{-2} at time t seconds.

(a) Find the velocity of P at time t seconds.
(b) Show that P is moving parallel to $\mathbf{i} + \mathbf{j}$ when $t = 4$.

> Note that the formulae for constant acceleration cannot be used here.
>
> $3 \times t^{\frac{1}{2}} \times \frac{2}{3} = 2t^{\frac{3}{2}}$.

(a) $\mathbf{v} = \int \mathbf{a}\, dt = \int (2t\mathbf{i} + 3\sqrt{t}\mathbf{j})dt$

$\qquad = t^2\mathbf{i} + 2t^{\frac{3}{2}}\mathbf{j} + \mathbf{c}$

When $t = 0$, $\mathbf{v} = 0 \Rightarrow \mathbf{c} = 0$ so, $\mathbf{v} = t^2\mathbf{i} + 2t^{\frac{3}{2}}\mathbf{j}$

(b) When $t = 4$, $\mathbf{v} = 16\mathbf{i} + 16\mathbf{j}$
$\qquad\qquad\qquad = 16(\mathbf{i} + \mathbf{j})$ so P is moving parallel to $\mathbf{i} + \mathbf{j}$.

Example

A force $(2t\mathbf{i} - 5\mathbf{j})$ N acts on a particle of mass 0.5 kg at time t seconds. Find the velocity of the particle at time t given that, initially, the velocity is $-3\mathbf{i} + 7\mathbf{j}$ m s^{-1}.

Using $\mathbf{F} = m\mathbf{a}$ gives $\qquad 2t\mathbf{i} - 5\mathbf{j} = 0.5\mathbf{a}$

$\Rightarrow \qquad\qquad\qquad\qquad \mathbf{a} = 4t\mathbf{i} - 10\mathbf{j}$

$\qquad\qquad\qquad\qquad \mathbf{v} = \int 4t\mathbf{i} - 10\mathbf{j}\, dt$

$\Rightarrow \qquad\qquad\qquad\qquad \mathbf{v} = 2t^2\mathbf{i} - 10t\mathbf{j} + c.$

When $t = 0$, $\mathbf{v} = -3\mathbf{i} + 7\mathbf{j} \Rightarrow \mathbf{c} = -3\mathbf{i} + 7\mathbf{j}$.

This gives $\qquad\qquad\qquad \mathbf{v} = 2t^2\mathbf{i} - 10t\mathbf{j} - 3\mathbf{i} + 7\mathbf{j}$

$\qquad\qquad\qquad\qquad \mathbf{v} = (2t^2 - 3)\mathbf{i} + (7 - 10t)\mathbf{j}.$

The velocity of the particle at time t is $(2t^2 - 3)\mathbf{i} + (7 - 10t)\mathbf{j}$ m s^{-1}.

Progress check

1 A small object is projected from ground level with speed 30 ms^{-1} at 40° above the horizontal. Take $g = 10$ ms^{-2}.

    (a) Find the speed and direction of the object after 1.5 seconds.

    (b) The object just clears a tree 25 m from the point of projection. How high is the tree?

2 A particle P of mass 0.4 kg moves in the direction of the positive x-axis with acceleration $\dfrac{5}{x}$ ms^{-2} away from O, where OP = x metres. When $x = 1$ the speed of P is 3 ms^{-1}. Find its speed when $x = 5$.

> Remember $F = ma$ Newton's 2nd law.

3 A particle P of mass 0.5 kg moves in a straight line, away from a point O, under the action of a force of magnitude $\dfrac{x^2}{12}$ N directed towards O, where OP = x metres. When $x = 3$ the speed of P is 15 ms^{-1}. Find x when P is brought to rest.

4 An object has position vector $\mathbf{r} = (2t^3\mathbf{i} - 5t^2\mathbf{j})$ m at time t seconds. Find the acceleration of the object when $t = 3$.

5 A particle P, initially at the point with position vector $(-3\mathbf{i} + 5\mathbf{j})$ m, has velocity $\mathbf{v} = \sqrt{t}\mathbf{i} + 3\mathbf{j}$ at time t seconds. Find the position vector of P when $t = 4$.

1 (a) 23.4 ms^{-1}, 10.6° above horiz (b) 15.2 m
2 5.01 ms^{-1} to 3 s.f.
3 12.7 m to 3 s.f.
4 $\ddot{\mathbf{r}} = 36\mathbf{i} - 10\mathbf{j}$
5 $\frac{3}{2}\mathbf{i} + 17\mathbf{j}$

3.2 Centre of mass

Centre of mass of a system of particles

In the diagram, m_1 and m_2 lie on a straight line through O. Their displacements from O are x_1 and x_2 respectively.

The position given by $\bar{x} = \dfrac{m_1 x_1 + m_2 x_2}{m_1 + m_2}$ is called the **centre of mass** of m_1 and m_2.

> If m_1 and m_2 are equal then this result gives the mid-point of the two masses.

For n separate masses m_1, m_2, ... m_n the position of the centre of mass is given by

$\bar{x} = \dfrac{m_1 x_1 + m_2 x_2 + ... m_n x_n}{m_1 + m_2 + ... m_n}$ This is usually written as $\bar{x} = \dfrac{\sum\limits_{i=1}^{n} m_i x_i}{\sum\limits_{i=1}^{n} m_i}$

In two dimensions, using the usual Cartesian coordinates, the position of the centre of mass is at $(\bar{x}, \bar{y})$ where $\bar{x}$ and $\bar{y}$ are given by

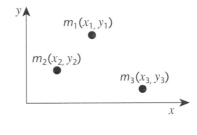

$\bar{x} = \dfrac{\sum\limits_{i=1}^{n} m_i x_i}{\sum\limits_{i=1}^{n} m_i}$ as above, and $\bar{y} = \dfrac{\sum\limits_{i=1}^{n} m_i y_i}{\sum\limits_{i=1}^{n} m_i}$

Example

Masses of 2 kg, 3 kg and 5 kg are positioned as shown in the diagram.
Find the coordinates of the centre of mass of the system.

> The method is easily extended to deal with any number of masses.

$\bar{x} = \dfrac{2 \times 2 + 3 \times 4 + 5 \times 7}{2 + 3 + 5} = \dfrac{51}{10} = 5.1$

$\bar{y} = \dfrac{2 \times 1 + 3 \times 5 + 5 \times 3}{2 + 3 + 5} = \dfrac{32}{10} = 3.2$

The centre of mass has coordinates (5.1, 3.2)

Centre of mass of a lamina

A **lamina** is something that is thin and flat such as a sheet of metal. The thickness of a lamina is ignored and it is treated as a two-dimensional object.

You can find the centre of mass of a uniform lamina by dividing it into parts which you then represent as particles. Each part is usually rectangular or triangular and in order to represent it as a particle:

- the mass of the particle is represented by the area of the part
- the position of the particle is taken to be at the geometric centre of the part.

This representation of a lamina gives you the information you need to apply the formulae for its centre of mass.

Example

Find the coordinates of the centre of mass of this uniform lamina.

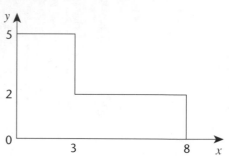

There are different ways to do this, but it doesn't matter which you choose.

The first step is to divide the lamina into rectangles.

Rectangle A has area 15 square units. Its centre is at (1.5, 2.5).

Rectangle B has area 10 square units. Its centre is at (5.5, 1).

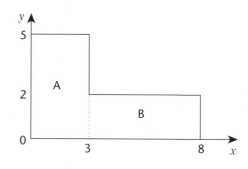

This process is easily extended to deal with more complex composite figures.

$$\bar{x} = \frac{15 \times 1.5 + 10 \times 5.5}{15 + 10} = 3.1$$

$$\bar{y} = \frac{15 \times 2.5 + 10 \times 1}{15 + 10} = 1.9$$

The centre of mass of the lamina has coordinates (3.1, 1.9).

Freely suspended lamina

Once you know where the centre of mass of a lamina is, you can use this to work out how it will move if it is freely suspended from a given point.

> When a lamina is freely suspended, it will move so that its centre of mass lies directly below the point of suspension.

KEY POINT

Example

The diagram shows the lamina of the previous example hanging freely from one corner.

Find the size of angle θ between the direction of Oy and the horizontal.

The position of the centre of mass is shown at the point G.
Notice that θ also lies in a right-angled triangle in which two of the sides are known.

From the diagram: $\tan \theta = \dfrac{1.9}{3.1}$

giving $\theta = 31.5°$ to 1 d.p.

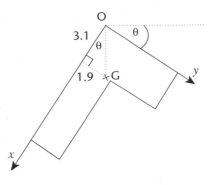

Progress check

1 Masses of 2 kg, 3 kg and 5 kg are positioned as shown in the diagram.

    (a) Find the coordinates of G, the centre of mass of the system.
    (b) Find the angle θ between OG and Ox.

1 (a) G(4.5, 2.5) (b) 29.1°

87

3.3 Work, energy and power

After studying this section you should be able to:

- understand kinetic and potential energy and the work energy principle
- understand the principle of conservation of mechanical energy
- understand the definition of power
- solve problems involving work, energy and power

LEARNING SUMMARY

Work and energy

In everyday language, **energy** is needed to do **work**. This idea is given a formal interpretation within mechanics.

The work done, W, by a *constant* force, F, in moving an object through a distance, d, along its line of action is given by $W = Fd$. The unit of work is the **joule** (J).

Energy is also measured in joules and the energy used in moving the object is equal to the work done. In a way, work and energy are like two sides of the same coin.

Energy may take different forms such as heat, sound, electrical, chemical and mechanical. An important property of energy is that it may change from one form to another. For your mechanics module, the focus of attention is on mechanical energy.

There are two forms of mechanical energy: **kinetic energy (KE)** and **potential energy (PE)**

The kinetic energy of an object is the energy it has due to its *motion*.

Its value depends on the mass of the object and its speed. $KE = \frac{1}{2}mv^2$.

For example, a mass of 4 kg moving with speed 5 ms^{-1} has $KE = \frac{1}{2} \times 4 \times 5^2 = 50$ J

The potential energy of an object is the energy it has due to its *position*.

> Potential energy may be gravitational or elastic. Elastic potential energy is covered in M3.

The **gravitational potential energy** of an object, relative to some level, is equal to the work done against gravity in raising the object from that level to its current position.

> The total mechanical energy of an object is the sum of its KE and PE.
>
> The **principle of conservation of mechanical energy** states that the total mechanical energy of a system remains constant, provided that it is not acted on by any external force, other than gravity.

KEY POINT

The principle of conservation of energy applies, for example, to the motion of a projectile. Any change in the PE of the projectile corresponds to an equal but opposite change in its KE.

Power

Power is the rate of doing work. It is measured in watts (W) where $1 \text{ W} = 1 \text{ Js}^{-1}$.

If the point of application of a force F moves with speed v in the *direction of the force* then the power P is given by $P = Fv$.

Example

Find the power generated by a car engine when the car travels at a constant speed of 72 kmh^{-1} on horizontal ground against resistance forces of 2500 N.

$$72 \text{ kmh}^{-1} = \frac{72 \times 1000}{60 \times 60} \text{ ms}^{-1} = 20 \text{ ms}^{-1}.$$

> Since the speed is constant, the force produced by the engine must match the resistance force.

$$\text{Power} = 2500 \times 20 \text{ W} = 50\,000 \text{ W}$$
$$= 50 \text{ kW}$$

Exam questions may involve resisted motion on an inclined plane.

Example

A car of mass 1000 kg travels up a hill inclined at angle θ to the horizontal, where $\sin \theta = \frac{1}{20}$. The non-gravitational resistance to motion is 2000 N and the power output from the engine is 60 kW. Find the acceleration of the car when it is travelling at 10 ms^{-1}. Take $g = 10 \text{ ms}^{-2}$.

> A clearly labelled diagram is a must.

> In this formula, F represents the *resultant force* up the hill.

Using $P = Fv$: $60\,000 = F \times 10 \Rightarrow F = 6000$

Using $F = ma$: $6000 - 2000 - 10\,000 \times \frac{1}{20} = 1000a$

This gives: $3500 = 1000a \Rightarrow a = 3.5$

The acceleration of the car up the hill is 3.5 ms^{-2}.

Progress check

1   A ball is kicked from ground level with speed 15 ms^{-1} and just clears a wall of height 2 m. The speed of the ball as it passes over the wall is $v \text{ ms}^{-1}$.

   (a) Assume that the ball has mass m kg and use the conservation of mechanical energy principle to write an equation.

   (b) Solve the equation to find v.

2   A car of mass 800 kg travels up a hill inclined at angle θ to the horizontal, where $\sin \theta = \frac{1}{20}$. The non-gravitational resistance to motion is 1600 N and the power output from the engine is 45 kW. Find the acceleration of the car when it is travelling at 15 ms^{-1}. Take $g = 10 \text{ ms}^{-2}$.

2   1.25 ms^{-2}
(b) 13.6 ms^{-1}
1   (a) $\frac{1}{2}m \times 15^2 = m \times 9.8 \times 2 + \frac{1}{2}mv^2$

3.4 Collisions and restitution

After studying this section you should be able to:

- *apply the principle of conservation of momentum to direct impact*
- *apply Newton's law of restitution using the coefficient of restitution*
- *calculate the loss of mechanical energy in a collision*
- *recall and use the definition of impulse as a change of momentum*

LEARNING SUMMARY

Conservation of momentum

> It's important to use the correct units.
> Remember that speed is the magnitude of the velocity.

The **momentum** of an object is the product of its mass and its velocity, i.e. momentum = $m\mathbf{v}$. Momentum has direction and so it is a vector quantity.

Momentum is measured in Ns (Newton seconds).

For example, an object of mass 5 kg moving with speed 6 ms^{-1} in a given direction has momentum $5 \times 6 = 30$ Ns in the direction of movement.

The **Principle of conservation of momentum** states that, in a collision, the total momentum before impact = the total momentum after impact.
Remember:

> **Key points from AS**
>
> - **Momentum**
>   *Revise AS page 72*

- always draw a diagram to represent the situation
- choose one direction to be positive (usually left to right $\longrightarrow$)
- show any unknown velocities in the positive direction.

Example

> When particles coalesce, they stick together and behave as a single particle.

Two particles A and B of masses 4 kg and 6 kg respectively move towards each other along the same straight line. Particle A is moving with speed 3 ms^{-1} and particle B is moving with speed 5 ms^{-1}. Given that the particles coalesce on impact, find their common velocity immediately afterwards.

> The velocity of B is in the negative direction.

Before impact:

After impact:

By the principle of conservation of momentum:

$$4 \times 3 - 6 \times 5 = 10v$$

giving: $\qquad 10v = -18$

so: $\qquad v = -1.8$

> v is negative so the direction of the velocity is from right to left.

The particles move with velocity 1.8 ms^{-1} in the direction from B to A.

Newton's law of restitution

Some problems require two unknown values to be found and in this situation a second equation is needed.

Newton found by experiment that, when two objects collide, the ratio of their speed of separation to their speed of approach has a fixed value. The symbol used to represent this ratio is e and its value is depends on what the objects are made of.

$$e = \frac{\text{speed of separation}}{\text{speed of approach}}$$

This is known as **Newton's law of restitution** and is often used to provide a second equation.

This is often used in the form $e \times$ speed of approach = speed of separation.

The speed of separation is always less than or equal to the speed of approach so it follows that $0 \leqslant e \leqslant 1$. The ratio e is called the **coefficient of restitution**.

Example

In the exam, you may need to deal with more than one impact. See the sample exam questions page 96.

A smooth sphere P moves towards a stationary smooth sphere Q in a straight line through their line of centres. P has mass 2 kg and speed 5 ms^{-1}. Q has mass 4 kg. Given that the coefficient of restitution between the spheres is 0.5, find:

(a) the speed of each sphere immediately after impact
(b) the loss of energy in the collision.

(a)

Before impact:

5 ms^{-1}

P (2 kg)   Q (4 kg)

After impact:

Momentum is conserved:
$2 \times 5 + 4 \times 0 = 2v_1 + 4v_2$

v_1 v_2

So:
$$v_1 + 2v_2 = 5 \tag{1}$$

Newton's law:
$$0.5 \times 5 = v_2 - v_1 \tag{2}$$

Adding (1) and (2) gives:
$$3v_2 = 7.5 \implies v_2 = 2.5$$
and:
$$v_1 = 0$$

Immediately after the impact, P is brought to rest and Q moves with speed 2.5 ms^{-1}.

Kinetic energy = $\frac{1}{2}mv^2$. Even though momentum is conserved, energy is lost.

(b) Loss of energy = KE before impact − KE after impact
$$= 0.5 \times 2 \times 5^2 - 0.5 \times 4 \times 2.5^2$$
$$= 12.5 \text{ J}$$

Impulse

Key points from AS

• **Impulse**
  Revise AS page 73

Impulse = change in momentum

When two particles collide, each receives an impulse from the other of equal size but opposite sign. In this way, the total change in momentum is zero. We would expect this to happen because the total momentum of the system is conserved in a collision.

Example

A ball of mass 0.5 kg moving with velocity $6\mathbf{i} + 2\mathbf{j}$ ms^{-1} is kicked and receives an impulse of $2\mathbf{i} + 11\mathbf{j}$ Ns. Find the velocity of the ball immediately after it is kicked.

Momentum immediately after kick = initial momentum + impulse

$$= 0.5(6\mathbf{i} + 2\mathbf{j}) + 2\mathbf{i} + 11\mathbf{j} = 5\mathbf{i} + 12\mathbf{j}$$
$$= 0.5\mathbf{v} \text{ (where } \mathbf{v} \text{ is the velocity)}$$

This gives:
$$\mathbf{v} = 10\mathbf{i} + 24\mathbf{j} \text{ ms}^{-1}$$

Progress check

1   Two particles A and B move towards each other with speeds of 6 ms^{-1} and 2 ms^{-1} respectively. A has mass 3 kg and B has mass 1 kg. The coefficient of restitution between the particles is 0.6. Find the speed of each particle after the collision.

1 A 2.8 ms^{-1}, B 7.6 ms^{-1}.

3.5 Statics of rigid bodies

Equilibrium

Forces that act in the same plane are coplanar.

You may need to show that a rigid body is in **equilibrium** under a system of **coplanar** forces *or* to *use* the fact that a rigid body is in equilibrium to find something about the forces acting on it.

To show that a rigid body is in equilibrium under the action of coplanar forces you need to establish that:

- the vector sum of the forces is zero
- the sum of the moments about some point is zero.

You only need to establish this for *one* point and you can choose any point that is convenient.

If a body is in equilibrium under some coplanar forces then you know that:

- the vector sum of the forces will be zero
- the sum of the moments about any point you choose will be zero.

> The conditions for equilibrium can be used to produce equations. The equations can then be solved to find the unknown values in a problem.
>
> KEY POINT

Example

A uniform plank AB of length 5 m and mass 20 kg rests horizontally on two supports, one at each end. A mass of 10 kg is positioned on the plank 1 m from B. Find the reaction at each of the supports. Take $g = 9.8$ ms^{-2}.

The first step is to draw a clearly labelled diagram.

The plank is uniform so its weight acts at its mid-point.

The vector sum of the forces is zero so the total upward force must equal the total downward force.

Vertically:
$$\mathbf{R}_A + \mathbf{R}_B = 20g + 10g$$
$$= 30g$$

M(A) gives:
$$5\mathbf{R}_B = 20g \times 2.5 + 10g \times 4$$
$$= 90g$$
$$\mathbf{R}_B = 18g$$
$$= 176.4 \text{ N (reaction at B)}$$

Substitute for $\mathbf{R}_B$:
$$\mathbf{R}_A + 18g = 30g$$
$$\mathbf{R}_A = 12g$$
$$= 117.6 \text{ N (reaction at A)}.$$

Leaning ladders

The situation where a ladder leans against a wall, possibly supporting a load at some point along its length, provides a rich source of questions on equilibrium. It requires forces and moments to be considered and includes the theory of friction.

Example

A uniform ladder of mass m kg rests against a smooth vertical wall with its lower end on rough horizontal ground. A man of mass $4m$ kg has climbed three-quarters of the way up the ladder but cannot go any higher or it will slip. The ladder makes an angle of 60° with the horizontal.

Find the coefficient of friction between the ladder and the ground.

Start by drawing a clearly labelled diagram.

> The wall is smooth so there is no vertical force at the top of the ladder.

> The length of the ladder has been written as $4l$ to simplify taking moments.

> Friction acts at the bottom of the ladder to oppose the tendency for it to slip.

In this example, three equations may be formed by resolving the forces in two directions and taking moments.

The equations provide enough information to solve the problem.

Vertically: $$\mathbf{R}_A = mg + 4mg \qquad (1)$$
$$= 5mg$$

Horizontally: $$F = \mathbf{R}_B \qquad (2)$$

M(A): $$\mathbf{R}_B \times 4l \sin 60° = mg \times 2l \cos 60° + 4mg \times 3l \cos 60°$$

giving: $$\mathbf{R}_B \times 2\sqrt{3}l = mgl + 6mgl$$

so: $$\mathbf{R}_B = \frac{7mg}{2\sqrt{3}} \qquad (3)$$

From (2) and (3) $$F = \frac{7mg}{2\sqrt{3}}$$

> It's a good idea to rationalise the denominator by multiplying the top and bottom of the fraction by $\sqrt{3}$.

Since the ladder is on the point of slipping, friction is limiting and the coefficient of friction is given by:

$$\mu = \frac{F}{\mathbf{R}_A} = \frac{7}{10\sqrt{3}} = \frac{7\sqrt{3}}{30}$$

Progress check

1 A uniform plank AB of length 3 m and mass 25 kg rests horizontally on two supports, one at each end. A mass of 15 kg is positioned on the plank 1 m from B. Find the reaction at each of the supports. Take $g = 9.8$ ms^{-2}.

2 A uniform ladder of length 5 m rests against a smooth wall with its base on rough horizontal ground. The coefficient of friction between the ladder and the ground is $\frac{2}{3}$ and the ladder is on the point of slipping. Show that the angle between the ladder and the ground is $\tan^{-1}(\frac{3}{4})$.

1 220.5 N (A), 171.5 N (B)

Sample questions and model answers

1

A ball of mass 250 g is released from rest 2 m above the ground. The coefficient of restitution between the ball and the ground is 0.7. Take $g = 9.8$ ms^{-2}.

(a) Find the height reached by the ball as it rebounds from the ground.

(b) Find the impulse exerted on the ball by the ground.

(a) Using $v^2 = u^2 + 2as$

gives $v^2 = 2 \times 9.8 \times 2$

$= 39.2$

so $v = \sqrt{39.2}$

There is no need to work out $\sqrt{39.2}$ at this stage.

The ball strikes the floor with speed $\sqrt{39.2}$ ms^{-1}.

Using Newton's law, the ball rebounds with speed $\sqrt{39.2} \times 0.7$ ms^{-1}

Some simple statements help to make your method clear.

Using $v^2 = u^2 + 2as$

gives $0 = 39.2 \times 0.49 - 2 \times 9.8 \times h$

so $h = \dfrac{39.2 \times 0.49}{2 \times 9.8} = 0.98$

The ball reaches a height of 0.98 m when it rebounds from the ground.

(b) Taking upwards as positive:

Momentum of ball immediately before it hits the ground

$= -0.25 \times \sqrt{39.2}$ Ns

Write the mass in kg.

Momentum of ball immediately after it hits the ground

$= 0.25 \times \sqrt{39.2} \times 0.7$ Ns

$-0.25 \times \sqrt{39.2}$ Ns $0.25 \times \sqrt{39.2} \times 0.7$ Ns

Impulse = change in momentum

$= 0.25 \times \sqrt{39.2} \times 0.7 - (-0.25 \times \sqrt{39.2})$

$= 2.66$ Ns to 3 s.f.

The impulse exerted on the ball by the ground is 2.66 Ns.

Sample questions and model answers (continued)

2

A particle P of mass $3m$ is moving with speed $2u$ when it strikes a particle Q of mass m which is at rest. The coefficient of restitution between the particles is e.

(a) Show that the speed of P after the collision is $\frac{v}{2}(3-e)$ and find the speed of Q.

(b) Q subsequently strikes a wall perpendicular to its direction of motion and rebounds so that a second collision with P occurs. Given that the coefficient of restitution between Q and the wall is $\frac{1}{3}$ find the speed of P after the second collision.

(c) Show that P moves in the same direction after the second collision.

(a) Before impact

After impact

Conservation of momentum gives: $6mu = 3mv_P + mv_Q$

which simplifies to $\qquad 6u = 3v_P + v_Q \qquad (1)$

law of restitution gives: $\qquad 2ue = v_Q - v_P \qquad (2)$

(1) − (2) gives: $\qquad 6u - 2ue = 4v_P$

giving $\qquad v_P = \frac{u}{2}(3-e)$ as required

> The approach is exactly the same as when numerical values of the speeds are given.

> This is the result given in the question.

and from (2) $\qquad v_Q = 2ue + v_P = 2ue + \frac{u}{2}(3-e)$

$$= \frac{u}{2}(4e + 3 - e) = \frac{3u}{2}(e+1)$$

The speed of Q after the collision is $\frac{3u}{2}(e+1)$.

(b) The speed of Q after hitting the wall is $\frac{1}{3} \times \frac{3u}{2}(e+1) = \frac{u}{2}(e+1)$ in the opposite direction.

Before impact

After impact

Conservation of momentum gives: $\frac{3mu}{2}(3-e) - \frac{mu}{2}(e+1) = 3mw_P + mw_Q$

which simplifies to $\qquad 4u - 2ue = 3w_P + w_Q \qquad (3)$

Newton's experimental law gives: $\left(\frac{u}{2}(3-e) + \frac{u}{2}(e+1)\right)e = w_Q - w_P$

which simplifies to $\qquad 2u = w_Q - w_P \qquad (4)$

(3) − (4) gives $\qquad 2u - 2ue = 4w_P$

and $\qquad w_P = \frac{u}{2}(1-e)$

> Repeat the process with the new speeds. Draw a diagram and set your working out clearly.

> There is no need to find w_Q.

so the speed of P immediately after the second collision is $w_P = \frac{u}{2}(1-e)$.

(c) $1 - e \geqslant 0$ since $0 \leqslant e \leqslant 1$ so the direction of movement of P is not changed.

Sample questions and model answers (continued)

3

A particle moving in a straight line passes through a fixed point O when $t = 0$. Its velocity at time t seconds is given by $v = 12 - 3t^2$ m s^{-1}.

(a) Find the acceleration of the particle when $t = 4$.

(b) Find the distance of the particle from O:

   (i)   when it comes to rest
   (ii) when $t = 5$.

(a) $v = 12 - 3t^2 \implies a = \dfrac{dv}{dt} = -6t$

   $t = 4 \implies a = -6 \times 4 = -24$.

The acceleration of the particle when $t = 4$ is -24 m s^{-2}.

(b) (i)   The particle comes to rest when $v = 0$.

   $v = 0 \implies 12 - 3t^2 = 0$

   $\implies t = 2$ since $t > 0$.

   The displacement of the particle from O at time t is given by

   $$x = \int v \, dt + c = \int 12 - 3t^2 \, dt + c$$

   $\implies x = 12t - t^3 + c$.

   When $t = 0$, $x = 0$

   giving $0 = 0 + c \implies c = 0$.

   So $x = 12t - t^3$.

   When $t = 2$ $x = 12 \times 2 - 2^3 = 16$.

   The particle is 16 m from O when it comes to rest.

(b) (ii) When $t = 5$ $x = 12 \times 5 - 5^3 = -65$.

   The particle is 65 m from O when $t = 5$.

Distance is taken to be positive.

Practice examination questions

1 A ball is struck with speed 25 ms^{-1} from a point 50 cm above the ground. When it has travelled 20 m horizontally, the ball just clears a fence that is 3 m high.

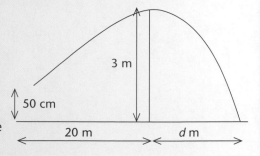

(a) Show that a possible angle of projection of the ball is 16.5°.

(b) Find the distance d m, between the fence and the point where the ball strikes the ground for this angle of projection.

2 A particle moves along the x-axis such that at time t seconds its displacement from a point O on the line is $x = t^3 - 6t^2 + 3$ metres.

(a) Show that the particle is initially at rest and find its greatest distance from O in the negative direction.

(b) Find the speed of P when it passes through its initial position.

3 The position vector of a particle P at time t seconds relative to a fixed origin O is given by $\mathbf{r} = (4t^2 + 5)\mathbf{i} + 2t^3\mathbf{j}$ metres.

Find:

(a) the speed of P when $t = 1$

(b) the time when P is moving parallel to the vector $\mathbf{i} + 3\mathbf{j}$.

4 The diagram shows a uniform lamina ABCDEFGH.

(a) Write down the distance of the centre of mass of the lamina from AH.

(b) Find the distance of the centre of mass of the lamina from AB.

(c) The lamina is now suspended freely from A. Find the angle that AB makes with the horizontal in its equilibrium position.

Practice examination questions (continued)

5

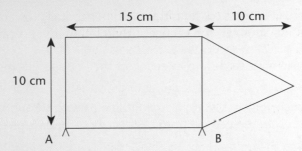

The diagram shows a uniform lamina of mass 0.8 kg resting on supports at A and B.

(a) Find the horizontal distance of the centre of mass from the support at A.

(b) Find the magnitude of the reaction force on the lamina at each support.

6

A and B are two smooth spheres of equal size and they are approaching each other on a direct line through their centres. A is moving with speed 4 ms^{-1} and B is moving with speed 2 ms^{-1}. The coefficient of restitution between the spheres is $\frac{1}{3}$.

Find:

(a) the speed of each sphere after the collision

(b) the impulse exerted by A on B during the collision

(c) the loss of mechanical energy during the collision.

7 Two particles P and Q approach each other with speeds u and $2u$ respectively. P has mass $3m$ and Q has mass m. The coefficient of restitution between the particles is e.

Given that the direction of motion of P is reversed in the collision, show that $e > \frac{1}{3}$.

8 A particle P of mass m is attached to one end of a light inelastic string of length l. The other end of the string is attached to a fixed point O. P is released from rest with the string taut and horizontal.

(a) Explain why, in the subsequent motion, the tension in the string does no work on the particle.

(b) Explain why it is not valid to use the formula $v^2 = u^2 + 2as$ in this situation.

(c) Find the maximum speed of the particle assuming that there is no resistance to motion.

9 A car of mass 900 kg travels up a straight road inclined at angle α to the horizontal, where $\sin \alpha = \frac{1}{15}$. Assume that the total non-gravitational force opposing the car's motion has a constant value of 800 N. Take $g = 9.8$ ms^{-2}.

(a) Find the driving force required for the car to maintain a steady speed up the slope.

(b) Find the power output of the car's engine given that its speed up the slope has a constant value of 20 ms^{-1}.

(c) At the top of the hill, the road surface is horizontal. If the power output of the engine remains the same, what is the initial acceleration of the car?

10 A football of mass 0.2 kg is travelling horizontally towards the goal with speed 20 ms^{-1} when it is struck by the keeper. It then rebounds vertically with speed 5 ms^{-1}.

Find the magnitude and direction of the impulse given to the ball by the keeper.

11 A particle P moves along a straight line passing through the point O. The displacement x metres, of the particle from O at time t seconds is given by:

$x = 4t^3 - 3t^2 + 7$ for $0 \leqslant t \leqslant 2$.

Find:

(a) the velocity of the particle at time t

(b) the distance that the particle travels before it changes direction

(c) the time taken for the particle to return to its starting point.

Chapter 4

Decision Mathematics 2

The following topics are covered in this chapter:

- *Linear programming*
- *The transportation problem*
- *Matching and allocation*

- *Game theory*
- *Networks*
- *Dynamic programming*

4.1 Linear programming

After studying this section you should be able to:

- formulate a linear programming problem in terms of decision variables
- use a graphical method to represent the constraints and solve the problem
- use the Simplex algorithm to solve the problem algebraically

LEARNING SUMMARY

Formulating a linear programming problem

To formulate a linear programming problem you need to:

x and y are often used for the variables.

- Identify the **variables** in the problem and give each one a label.
- Express the **constraints** of the problem in terms of the variables. You need to include non-negativity constraints such as $x \geqslant 0$, $y \geqslant 0$.

Typically, this may be to maximise a profit or minimise a loss.

- Express the quantity to be optimised in terms of the variables. The expression produced is called the **objective function**.

Example
A small company produces two types of armchair. The cost of labour and materials for the two types is shown in the table.

	Labour	Materials
Standard	£30	£25
Deluxe	£40	£50

The total spent on labour must not be more than £1150 and the total spent on materials must not be more than £1250. The profit on a standard chair is £70 and the profit on a deluxe chair is £100. How many chairs of each type should be made to maximise the profit?

In this case, the variables are the number of chairs of each type that may be produced. Using x to represent the number of standard chairs and y to represent the number of deluxe chairs, the constraints may be written as:

$$30x + 40y \leqslant 1150 \Rightarrow 3x + 4y \leqslant 115$$
$$25x + 50y \leqslant 1250 \Rightarrow x + 2y \leqslant 50$$

It's a good idea to simplify the constraints where possible.

and $\qquad x \geqslant 0, \quad y \geqslant 0.$

Using P to stand for the profit, the problem is to maximise $P = 70x + 100y$.

The graphical method of solution

Each constraint is represented by a region on the graph.
It's a good idea to shade out the *unwanted* region for each one. The part that remains unshaded then defines the **feasible region** containing the points that satisfy all of the constraints.

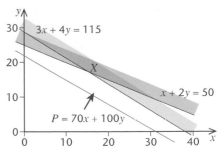

The blue line represents the points where the profit takes a particular value. Moving the line in the direction of the arrow corresponds to increasing the profit. This suggests that the maximum profit occurs at the point X.

Solving $3x + 4y = 115$ and $x + 2y = 50$ simultaneously gives X as (15, 17.5).

> X does not represent the solution in this case because both x and y must be integers.

The nearest points with integer coordinates in the feasible region are (15,17) and (14,18) The profit, given by $P = 70x + 100y$, is greater at (14,18).

The maximum profit is made by producing 14 standard and 18 deluxe chairs.

The Simplex algorithm

When there are three or more variables, a different approach is needed. The **Simplex algorithm** may be used to solve the problem algebraically. The information must be expressed in the right form before the algorithm can be used.

- First write the constraints, other than the non-negativity conditions, in the form $ax + by + cz \leqslant d$.
- Write the objective function in a form which is to be maximised.
- Add **slack variables** to convert the inequalities into equations.
- Write the information in a table called the **initial tableau**.

> The objective function is already in a form which is to be maximised.

Example

Find the maximum value of $P = 2x + 3y + 4z$ subject to the constraints:

$$3x + 2y + z \leqslant 10 \qquad \text{[i]}$$
$$2x + 5y + 3z \leqslant 15 \qquad \text{[ii]}$$
$$x \geqslant 0, \quad y \geqslant 0, z \geqslant 0.$$

> The constraints are in the right form.

Using slack variables s and t, the inequalities [i] and [ii] become:

$$3x + 2y + z + s = 10$$
$$2x + 5y + 3z + t = 15.$$

The objective function must be rearranged so that all of the terms are on one side of the equation to give:

$$P - 2x - 3y - 4z = 0.$$

The information may now be put into the initial tableau.

	P	x	y	z	s	t	value
The top row shows the objective function	1	−2	−3	−4	0	0	0
This row shows the first constraint	0	3	2	1	1	0	10
This row shows the second constraint	0	2	5	3	0	1	15

> You need to know how to read values from the tableau.

The columns for P, s and t contain zeros in every row apart from one. In each case, the remaining row contains 1 and the value of the variable is given in the end column of that row. The value of every other variable is taken to be zero. So, at this stage, the tableau shows that $P = 0$, $s = 10$, $t = 15$ and x, y and z are all zero. This corresponds to the situation at the origin.

Using the algorithm is equivalent to visiting each vertex of the feasible region until an optimum solution is found. This occurs when there are no negative values in the objective row.

When the present tableau is not optimal, a new tableau is formed as follows:

- The column containing the most negative value in the objective row becomes the **pivotal column**. In this case the pivotal column corresponds to the variable z.
- Now divide each entry in the *value* column by the corresponding entry in the pivotal column provided that the pivotal column entry is positive.
  For the example above this gives $\frac{10}{1} = 10$ and $\frac{15}{3} = 5$
  The smallest of these results relates to the bottom row which is now taken to be the **pivotal row**. The entry lying in both the pivotal column and the pivotal row then becomes the pivot. In this case, the pivot is 3.

P	x	y	z	s	t	value	
1	−2	−3	−4	0	0	0	
0	3	2	1	1	0	10	
0	2	5	3	0	1	15	← pivotal row

pivotal column

- Divide every value in the pivotal row by the pivot, this makes the value of the pivot 1. The convention is to use fraction notation rather than decimals.

P	x	y	z	s	t	value	
1	−2	−3	−4	0	0	0	
0	3	2	1	1	0	10	
0	$\frac{2}{3}$	$\frac{5}{3}$	1	0	$\frac{1}{3}$	5	← pivotal row

- Now turn the other values in the pivotal column into zeros by adding or subtracting multiples of the pivotal row.

See paragraph on preceding page about reading values from the tableau.

P	x	y	z	s	t	value	
1	$\frac{2}{3}$	$\frac{11}{3}$	0	0	$\frac{4}{3}$	20	
0	$\frac{7}{3}$	$\frac{1}{3}$	0	1	$-\frac{1}{3}$	5	
0	$\frac{2}{3}$	$\frac{5}{3}$	1	0	$\frac{1}{3}$	5	← pivotal row

It is often necessary to repeat the process in order to find the optimum tableau.

- If there are no negative values in the objective row then the **optimum tableau** has been found. Otherwise choose a new pivotal column and repeat the process.

Each time through the process is called an **iteration**.

In this case the optimum tableau has been found after just one iteration.
It shows that the maximum value of P is 20 and that this occurs when

$x = 0$, $y = 0$, $z = 5$, $s = 5$ and $t = 0$.

Progress check

An initial simplex tableau for a linear programming problem is given by:

P	x	y	z	s	t	value
1	−3	−2	−1	0	0	0
0	2	3	1	1	0	10
0	3	4	2	0	1	18

1 Carry out one iteration to produce the next tableau.

2 Explain why the tableau found is optimal. Write down the maximum value of P and the corresponding values of the other variables.

1

P	x	y	z	s	t	value
1	0	$\frac{5}{2}$	$\frac{1}{2}$	$\frac{3}{2}$	0	15
0	1	$\frac{3}{2}$	$\frac{1}{2}$	$\frac{1}{2}$	0	5
0	0	$-\frac{1}{2}$	$\frac{1}{2}$	$-\frac{3}{2}$	1	3

2 There are no negative values in the objective row.
$P = 15$, $x = 5$, $y = 0$, $z = 0$, $s = 0$, $t = 3$.

4.2 The transportation problem

After studying this section you should be able to:

- understand the transportation problem and how it can be formulated as a linear programming problem
- use the North-West corner rule to find an initial feasible solution
- use the stepping-stone method with improvement indices for obtaining an optimal solution

LEARNING SUMMARY

The transportation problem

The **transportation problem** deals with the distribution of goods from several **sources** of supply, such as factories, to a number of **destinations**, such as warehouses. Each source has a fixed **capacity** (that is, it is able to supply a fixed number of units) and each destination has a fixed **demand** (that is, it requires a fixed number of units). The objective is to schedule the movement of units from sources to destinations so as to minimise transport costs.

Factory	Capacity
A	3
B	5
C	7
Total supply	15

Warehouse	Requirement
P	4
Q	6
R	5
Total demand	15

COSTS			
From \ To	P	Q	R
A	6	5	4
B	7	6	7
C	8	7	9

To solve a transportation problem you should:

1. Set up a transport table and fill in the supply, demand and cost figures. Check that supply equals demand. If it does not, the problem is **unbalanced** and you should insert a dummy row or column with zero costs to take up the slack.

From \ To	P	Q	R	Total supply
A	6	5	4	3
B	7	6	7	5
C	8	7	9	7
Total demand	4	6	5	15

In the optimal solution the slack will be allocated to the dummy and will represent unmet demand or supply.

2. Find an initial solution using **the NW corner rule**. STEP 1: Start in the left hand cell of row 1 (NW corner) and fill in the maximum number of units which can move along this route by considering the row

From \ To	P	Q	R	Total supply
A	3 (6)	5	4	3
B	1 (7)	4 (6)	7	5
C	8	2 (7)	5 (9)	7
Total demand	4	6	5	15

and column totals. STEP 2: Repeat for each cell in row 1 until the supply for that row is exhausted. STEP 3: Repeat steps 1 and 2 for subsequent rows until all supply and demand is exhausted. Calculate your initial cost: (3×6) + (1×7) + (4×6) + (2×7) + (5×9) = 108

3. Check that the number of occupied cells equals rows + columns – 1. If not, the problem is degenerate. In this case place a zero in the unoccupied cell with the lowest cost and proceed, treating this cell as though it is occupied.

4. Calculate the **improvement indices** for each unoccupied cell.
   - If they are all greater than or equal to zero you have the optimal solution. Stop and calculate the optimal cost.
   - If one or more of the indices are negative, use the stepping-stone method to move as many units as possible along the cycle that includes the biggest negative index.

5. Repeat steps 3 and 4 until you have an optimal solution.

Calculating improvement indices

First calculate the row values (R_i) and column values (K_j). To do this let $R_1 = 0$ and find the other Rs and Ks from the table using the costs in the **occupied** cells only and the fact that $R_i + K_j = C_{ij}$ (where C_{ij} is the cost for row i, column j). Calculate $I_{ij} = C_{ij} - R_i - K_j$ for each **unoccupied** cell (where I_{ij} is the improvement index for row i, column j)

From \ To	P	Q	R	Total supply	R_i
A	6 — 3	0 5 -3	4	3	0
B	7 — 1	6 — 4	-1 7	5	1
C	0 8	7 — 2	9 — 5	7	2
Total demand	4	6	5	15	
K_j	6	5	7		

Stepping-stone method

> A **cycle** is complete when the cell moved to is in the same row or column as the cell entered.

We want to use the route with the biggest negative improvement index. This cell will **enter** the solution. Let θ be the number of units through that route and find a **cycle** such that
- All cells in the cycle are occupied except the entering cell.
- Each step in the cycle is to a cell in the same row or column, alternating horizontal and vertical steps
- No cell is used more than once

Add and subtract θ to cells in the cycle so that all constraints are met. Maximise θ subject to the constraint that all cells must be non-negative. One cell will **exit** the solution.

From \ To	P	Q	R	Total supply
A	6 — $3-\theta$	5 — Enters $+\theta$	4	3
B	7 — $1+\theta$	6 — $4-\theta$	7	5
C	8	7 — $2+\theta$	9 — $5-\theta$	7
Total demand	4	6	5	15

From \ To	P	Q	R	Total supply
A	Exits 6	5	4 — 3	3
B	7 — 4	6 — 1	7	5
C	8	7 — 5	9 — 2	7
Total demand	4	6	5	15

Find cycle and add or subtract θ from cells so that all constraints are met

Maximise θ. In this case $\theta = 3$ and the improved cost is 99

Although we have an improved cost, we have not established whether or not this is optimal. In fact, repeating steps 3 and 4 shows that it is not optimal and the process must be gone through again to obtain an optimal cost of 98. In the exam, you may not need to find the optimal solution. If a cell in an optimal solution has an improvement index equal to zero, then an alternative optimal solution can be found using that cell.

Formulation as a linear program

The transportation problem can also be formulated as a linear program.
Let x_{ij} be the units transported from factory i (i = A, B, C) to warehouse (j = P, Q, R).
Then the problem becomes:
Minimise the total transportation cost Z where
$$Z = 6x_{AP} + 5x_{AQ} + 4x_{AR} + 7x_{BP} + 6x_{BQ} + 7x_{BR} + 8x_{CP} + 7x_{CQ} + 9x_{CR}$$
Subject to
supply constraints $\quad x_{AP} + x_{AQ} + x_{AR} = 3 \quad x_{BP} + x_{BQ} + x_{BR} = 5 \quad x_{CP} + x_{CQ} + x_{CR} = 7$
demand constraints $\quad x_{AP} + x_{BP} + x_{CP} = 4 \quad x_{AQ} + x_{BQ} + x_{CQ} = 6 \quad x_{AR} + x_{BR} + x_{CR} = 5$
and the non-negative requirement $x_{ij} \geqslant 0$

Progress check

1 A company wishes to transport a product from warehouses to shops with the following capacities, requirements and associated costs:

Warehouse	Capacity
Manchester	20 units
Birmingham	10 units
London	25 units
Total supply	55 units

Shop	Requirement
Manchester	11 units
Coventry	13 units
Birmingham	17 units
Cardiff	14 units
Total demand	55 units

From \ To	Manchester	Coventry	Birmingham	Cardiff
Manchester	1	6	3	6
Birmingham	7	3	1	6
London	9	4	5	4

Transportation costs (£ per unit)

Find the optimal solution and state its minimum cost.

2 A car hire company has decided to close two offices and transfer their cars to more profitable sites. The decision will be based on the cost of relocation. The costs involved ($ per car) are shown in the following table:

From \ To	San Francisco	Los Angeles	New York	Cars Available
Atlanta	28	32	40	50
Houston	36	28	32	60
Cars Required	70	30	40	

(a) Use the North-West corner rule to find an initial solution.
(b) Find the optimal solution and state the minimum cost of relocation.

1   Minimum cost is £152.

	Manchester	Coventry	Birmingham	Cardiff
Manchester	11		9	
Birmingham		2	8	
London		11		14

2   This problem is unbalanced.
Minimum cost is $3200.

	San Francisco	Los Angeles	New York	
Atlanta	50			
Houston		30	30	
Unmet demand	20		10	

4.3 Matching and allocation

After studying this section you should be able to:

- *use bipartite graphs to model matchings*
- *understand the conditions for matchings to be maximal or complete*
- *apply the maximum matching algorithm*

LEARNING SUMMARY

Matchings and graphs

A **bipartite graph** is a graph in which the vertices are divided into two sets such that no pair of vertices in the same set is connected by an edge.

In this case, the two sets are {A, B, C} and {p, q, r, s}.

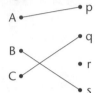

Some vertices in a bipartite graph may not be connected to another vertex.

A **matching** between two sets may be represented by a bipartite graph in which there is at most one edge connecting a pair of vertices.

A **maximal matching** is a matching which has the maximum number of edges. This occurs when every vertex in one of the sets is connected to a vertex in the other set. The bipartite graph shown above represents a maximal matching.

A **complete matching** is a matching in which every vertex is connected to another vertex. This can only occur when the two sets contain the same number of vertices.

The matching improvement algorithm

Figure 1 is a bipartite graph showing the possible connections between two sets. It does not represent a matching because some vertices have more than one connection.

Figure 2 is a bipartite graph representing an initial matching.

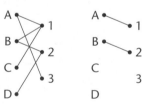

Figure 1   Figure 2

An initial matching may be improved by increasing the number of connections. This is the purpose of the **matching improvement algorithm**.

Alternating path   Maximal matching

> In an alternating path, the edges alternate between those from the bipartite graph that are not in the initial matching and those that are.

Step 1   In the initial matching, start from a vertex that is not connected and look for an **alternating path** to a vertex in the other set that is not connected.

Step 2   Each edge on the alternating path, not included in the initial matching, is now included and each edge originally included is removed.

Step 3   Repeat steps 1 and 2 using the latest matching in place of the initial matching until no further alternating paths can be found.

Allocation problems

In an allocation (or assignment) problem, the object is to set up a matching that will optimise a particular objective, such as minimising a cost or maximising a profit. If we regard one set of objects as agents and the other set as tasks, then the underlying assumptions are that:

- Any agent may be assigned to any task (unlike the matching problems).
- Each agent is assigned at most one task and each task has at most one agent assigned to it.

When the number of agents is equal to the number of tasks, the problem is said to be **balanced**. If the number of agents exceeds the number of tasks, or vice versa, then the problem is **unbalanced**. The technique for solving an unbalanced problem involves introducing a **dummy** agent or task which then allows the solution to proceed as for the balanced case. In the solution of an unbalanced problem, a task assigned to a dummy agent isn't actually completed and an agent assigned to a dummy task doesn't actually have a task to carry out.

The opportunity cost matrix

The table shows the time in hours for three workers A, B and C to complete three separate tasks.

		Task		
		1	2	3
Worker	A	9	11	7
	B	8	10	9
	C	7	9	8

The **cost matrix** shows only the figures in hours. In this case, the 'cost' is measured in time.

9	11	7
8	10	9
7	9	8

The **opportunity cost matrix** is found from the cost matrix by:

1. Subtracting the smallest value in each row from every other value in that row.
2. Subtracting the smallest value in each column from every other value in that column.

9	11	7
8	10	9
7	9	8

$\xrightarrow{1}$

2	4	0
0	2	1
0	2	1

$\downarrow 2$

2	2	0
0	0	1
0	0	1

The Hungarian algorithm

The **Hungarian algorithm** is used to find the allocation that optimises the given objective.

Step 1 Find the opportunity cost matrix.

Step 2 Test for optimality. If true then make the allocation and stop.

Step 3 Revise the opportunity cost matrix and repeat step 2.

109

The method used to test for optimality is to find the minimum number of horizontal and/or vertical lines needed to cover all of the zeros in the opportunity cost matrix. If this number = n where the matrix has n rows and n columns, then the matrix will give an optimal allocation.

In the example, the number of lines needed = 3.
The matrix has 3 rows and columns so it will give an optimal allocation.

2	2	0
0	0	1
0	0	1

The optimal allocation is now found by selecting positions of zeros in the opportunity cost matrix such that one zero is selected in each row and in each column.

One possibility is shown here along with the corresponding figures in the original cost matrix.

2	2	**0**
0	**0**	1
0	0	1

9	11	**7**
8	**10**	9
7	9	8

The allocation is A → 3, B → 2 and C → 1 giving a total cost, in hours, of 7 + 10 + 7 = 24.

In this case, there is an alternative solution.

2	2	**0**
0	0	1
0	**0**	1

9	11	**7**
8	10	9
7	**9**	8

The allocation is A → 3, B → 1 and C → 2 giving a total cost, in hours, of 7 + 8 + 9 = 24.

Revising the opportunity cost matrix

When the opportunity cost matrix does not give an optimal solution, a revised opportunity cost matrix is found as follows:

Step 1 Find the smallest number not covered by a line. Subtract this number from every number not covered by a line.

Step 2 Add this number to every number lying at the intersection of two lines.

For example, this opportunity cost matrix has 3 rows and columns, but its zeros can be covered with just 2 lines.

0	3	1
0	2	5
0	0	0

Applying the steps given above produces this revised opportunity cost matrix.

0	2	**0**
0	1	4
1	**0**	0

Maximising allocation problems

If the objective involves maximising a quantity, such as a profit, then the first step is to subtract every value in the original matrix from a fixed value. This fixed value may be taken to be the largest value in the matrix or any value larger than this. The Hungarian algorithm is now applied in the usual way to this transformed matrix.

Progress check

1 Starting with an initial matching in which 1 is connected to R and 2 is connected to P, use the improvement algorithm to establish a complete matching for this bipartite graph.

2 Given the cost matrix:

	1	2	3
A	3	3	4
B	8	5	9
C	2	1	4

Use the Hungarian algorithm to find the minimum total cost.

1 1 – P, 2 – S, 3 – R, 4 – Q
2 4 + 5 + 2 = 11

111

4.4 Game theory

After studying this section you should be able to:

- understand the idea of a zero-sum game and its representation using a pay-off matrix
- identify play-safe strategies and stable solutions
- find optimal mixed strategies for a game with no stable solution
- reduce a pay-off matrix using a dominance argument

LEARNING SUMMARY

A two person zero-sum game

The starting point in the mathematical theory of games is that the outcome of a game is determined by the **strategies** of the players.

A two person **zero-sum game** is a game in which the winnings of one player equal the losses of the other for every combination of strategies. Taking winnings to be positive and losses to be negative gives a zero sum in each case.

Viewing a game from one player's point of view, we could represent the outcomes (called pay-offs) for each combination of strategies in a matrix. This is called the **pay-off matrix** for that player.

Example

A and B are two players in a zero-sum game. A uses one of two strategies, W or X, and B uses one of the strategies Y or Z. The table shows the pay-off matrix for A.

> The situation could equally be represented by the pay-off matrix for B. This would show corresponding values with opposite signs since this is a zero-sum game.

Pay-off matrix for A		B	
		Y	Z
A	W	2	−2
	X	5	−4

The pay-off matrix shows that if B adopts strategy Y then the pay-off for A will be 2 by using strategy W and 5 using strategy X. On the other hand, if B adopts strategy Z then the pay-off for A will be −2 using strategy W and −4 using strategy X.

The idea is that neither player knows in advance which strategy the other will use.

The play-safe strategy

The **play-safe** strategy for a player is the strategy for which the minimum pay-off is as high as possible. In the example above, the minimum pay-off for A using strategy W is −2, whereas the minimum pay-off using strategy X is −4. This means that strategy W is the play-safe strategy for player A.

Notice that finding the play-safe strategy for player A involves comparing the minimum values in the *rows* of the pay-off matrix for A. Finding the play-safe strategy for B will involve comparing the values in the *columns*, remembering that B's pay-offs are the negatives of the ones in the pay-off matrix for A.

The minimum value for B using strategy Y is −5 and the minimum value using strategy Z is 2. This means that the play-safe strategy for B is strategy Z.

The situation is shown in the pay-off matrices for A and B.

Pay-off matrix for A		B	
		Y	Z
A	W	2	−2
	X	5	−4

Pay-off matrix for B		B	
		Y	Z
A	W	−2	2
	X	−5	4

In this case, the maximum of the minimum pay-offs, for each player, is in the corresponding position in the two matrices. This represents the **stable solution** to the problem referred to as the **saddle point** (or **minimax point**).

The solution is stable in the sense that neither player can improve their pay-off by taking a different strategy, given that the other player doesn't change. In other words, while B uses strategy Z, the best strategy for A is W and while A uses strategy W, the best strategy for B is Z.

In the example, the sum of these values is −2 + 2 = 0 and the solution is stable.

> **KEY POINT**
> If the sum of the two values used to determine the play-safe strategies is **not zero** then the values cannot correspond to the same cell position in the play-off matrices. This means that there is no saddle point and the game has **no stable solution**.

Example

The pay-off matrices for two players in a zero-sum game are given below. Show that there is no stable solution for the game.

Pay-off matrix for P		Q	
		Y	Z
P	W	5	−3
	X	−4	6

Pay-off matrix for Q		Q	
		Y	Z
P	W	−5	3
	X	4	−6

The play-safe strategies for P and Q are shown shaded.
The values used to determine the play-safe strategies are −3 and −5.
There is no stable solution since $(-3) + (-5) \neq 0$.

Optimal strategies for games that are not stable

The repeated use of the same strategy over a series of games is called a **pure strategy**. This provides the best results for both players in a game which has a stable solution. In the case where no stable solution exists, a **mixed strategy** is used in which each of the strategies is employed with a given probability to find the optimal solution.

Returning to the previous example:

Pay-off matrix for P		Q	
		Y	Z
P	W	5	−3
	X	−4	6

Suppose that P chooses:

strategy W with probability p

and   strategy X with probability $(1 - p)$.

Then if:

Q chooses strategy Y, the expected *gain* for P is given by $5p - 4(1 - p) = 9p - 4$

Q chooses strategy Z, the expected *gain* for P is $-3p + 6(1 - p) = -9p + 6$

> The optimal value occurs when these expressions are equal.

$$9p - 4 = -9p + 6 \Rightarrow 18p = 10$$
$$\Rightarrow p = \tfrac{5}{9}$$

> You could equally use $-9p + 6$ to get the value of the game.

The value of the game is given by $9p - 4 = 1$.

Suppose that Q chooses:

strategy Y with probability q

and   strategy Z with probability $(1 - q)$.

Then if:

P chooses strategy W, the expected *loss* for Q is given by $5q - 3(1 - q) = 8q - 3$

P chooses strategy X, the expected *loss* for Q is $-4q + 6(1 - q) = -10q + 6$

$$8q - 3 = -10q + 6 \Rightarrow 18q = 9$$
$$\Rightarrow q = \tfrac{1}{2}$$

(As a check, the value of the game is $8q - 3 = 1$ as before).

This shows that the optimal strategy for both players is to use mixed strategies such that:

P chooses strategy W with probability $\tfrac{5}{9}$ and strategy X with probability $\tfrac{4}{9}$

Q chooses strategy Y and strategy Z with equal likelihood.

The expected long-term gain then for P, as an average per game, is 1 and this is also the expected long-term loss, as an average per game, for Q.

Graphical representation

Each graph corresponds to a strategy for Q (i.e. the opponent of P).

The diagram shows graphs of $9p - 4$ and $-9p + 6$ against values of p from 0 to 1.

The point of intersection corresponds to the probability that gives the optimal mixed strategy for P in the last example.

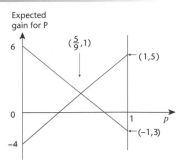

> **KEY POINT**
>
> When P's opponent has more strategies there will be more graphs with several points of intersection. You will need to identify the one that represents the optimal mixed strategy for P.
> *This will be the highest point on or below each of the graphs.*

This diagram represents a situation where P's opponent has three strategies to choose from.

The point representing the optimal mixed strategy for P is circled.

Notice how the problem of identifying the right vertex can be expressed as a linear programming problem in which the object is to maximise the expected gain for P subject to the constraints represented by the regions bounded by the straight line graphs.

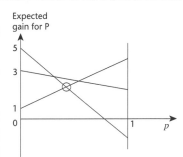

This is, in fact, the approach used for higher dimensional problems. The conditions are formulated as a linear programming problem which may then be solved by the simplex algorithm.

> **KEY POINT**
>
> If the pay-offs for one strategy are *always* better than the corresponding pay-offs for some other strategy then the weaker one can be ignored when determining the probabilities for a mixed strategy. In this way, the pay-off matrix is reduced by what is called a **dominance argument**.

Progress check

1 The table shows a pay-off matrix for player A in a zero-sum game.

(a) Find the play-safe strategy for each player.

(b) Show that this game has no stable solution.

(c) Find the best strategy for each player and the value of the game.

Pay-off matrix for A	Q	
	Y	Z
A W	6	−4
A X	−3	5

4.5 Networks

After studying this section you should be able to:

- *understand flows in networks and be able to find the maximum flow for a network involving multiple sources and sinks*
- *understand the use of the maximum flow – minimum cut theorem*
- *understand the travelling salesman problem and calculate upper and lower bounds for the total distance*

LEARNING SUMMARY

Flows in networks

The **flows** referred to may be flows of liquids, gases or any other measurable quantities. The edges may represent such things as pipes, wires or roads that carry the quantities between the points identified as vertices.

A typical vertex has a flow into it and a flow out of it. The exceptions are a **source** vertex which has no input and a **sink** vertex which has no output.

Each edge of the network has a **capacity** which represents the maximum possible flow along that edge. In the usual notation, the capacity is written next to the edge and the flow is shown in a circle.

This edge has a capacity of 10 and a flow of 7.

The set of flows for a network is **feasible** if:

- The total output from all source vertices is equal to the total input for all sink vertices.

- The input for each vertex other than a source or sink vertex is equal to its output.

- The flow along each edge is less than or equal to its capacity.

A **cut** divides the vertices into two sets, one set containing the source and the other containing the sink.

The **capacity of a cut** is equal to the sum of the capacities of the edges that cross the cut, *taken in the direction from the source set to the sink set.*
In the diagram, the capacity of cut (i) is 15 + 10 = 25.
The capacity of cut (ii) is 8 + 14 = 22.
Notice that, for the second cut, the capacity of 4 is not included because it does not go from the source set to the sink set.

> Be careful to add the capacities and not the flows.

The maximum flow–minimum cut theorem

The **maximum flow–minimum cut theorem** states that the maximum value of the flow through a network is equal to the capacity of the minimum cut. (This is a little bit like saying that a chain is only as strong as its weakest link.)

It follows from the theorem that if you find a flow that is equal to the capacity of some cut then it must be the maximum flow that can be established through the network.

For example, the capacity of cut (ii) for this network has already been found to be 22. The circled figures show one way of achieving this flow and the theorem now tells us that this must be the maximum possible flow through the network.

Flow augmentation

Once an initial flow has been established through a network it is useful to know the extent to which the flow along any edge may be altered in either direction.

This is labelled as **excess capacity** and **back capacity** on each edge as shown. It may be possible to **augment** the flow through the network by making use of this flexibility along a path from the source vertex to the sink vertex. Such a path is called a **flow-augmenting path**.

A **saturated** edge is one where the flow is equal to the capacity.

An algorithm for finding the maximum flow through a network by augmentation is:

Step 1   Find an initial flow by inspection.

Step 2   Label the excess capacity and back capacity for each edge.

Step 3   Search for a flow-augmenting path. If one can be found then increase the flow along the path by the maximum amount that remains feasible.

Step 4   Repeat steps 2 and 3 until no flow-augmenting path may be found.

Multiple sources and sinks

A network may have several sources S_1, S_2, ... and several sinks T_1, T_2, ... but the methods described in this section can still be applied through the use of a **supersource** and a **supersink**. A supersource S is an extra node, added to the network, that acts as a source for each of the original sources. Similarly, a supersink T is a node, added to the network, that acts as a sink for each of the original sinks.

Example

The network below shows two sources S_1 and S_2 and two sinks T_1 and T_2. Find the maximum flow through the network and state any flow-augmenting paths.

Explain why the flow is maximal.

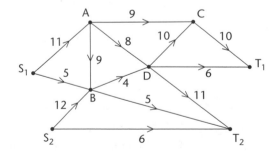

The first step is to add the supersource S and the supersink T to the network.

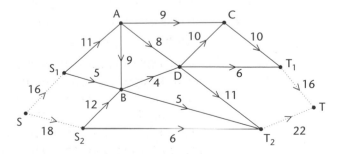

The usual convention is to use dotted lines for the edges linking S and T to the network.

The capacities on the edges SS_1 and SS_2 have been chosen to support the maximum possible flows from S_1 and S_2. Similarly, the capacities on the edges T_1T and T_2T have been chosen to support the maximum flows into T_1 and T_2. The diagram also shows a cut with capacity $11 + 4 + 5 + 6 = 26$.

A flow of 26 can be achieved as shown in the table.

Flow-augmenting path	Flow	Saturated edges
SS_1ACT_1T	9	S_1A, AC
SS_1ADT_1T	2	S_1A, DT_1
SS_2BDT_1T	4	BD, DT_1
SS_2BT_2T	5	BT_2
SS_2T_2T	6	S_2T_2
Total	26	

The flows on the flow-augmenting paths are shown circled on the network below. The flow of 26 is maximal since it equals the capacity of the cut (maximum flow minimum cut theorem).

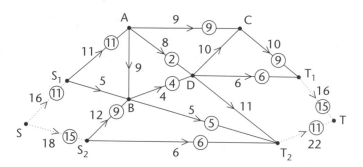

Note that each of the edges on the minimum cut is saturated. This will always be the case.

The final stage is to remove the supersource and supersink, along with the extra arcs, to show the maximum flow pattern for the original network.

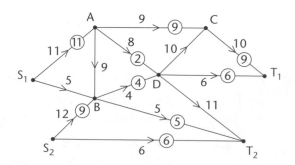

The travelling salesman problem

This is based on the classic situation of a salesman who wishes to visit a number of towns and return home using the shortest possible route.

The **travelling salesman problem** (TSP) is the problem of finding a route of minimum distance that visits every vertex and returns to the start vertex. For a small network it is possible to produce an exhaustive list of all possible routes and choose the one that minimises the total distance. For a large network an exhaustive check is not feasible, even with a computer, because the number of possible routes grows so rapidly.

It is useful to know within what limits the total distance must lie and there are algorithms that can be used for this purpose.

Finding an upper bound

> An upper bound for the total distance involved in the travelling salesman problem is given by twice the minimum spanning tree.

KEY POINT

The **minimum spanning tree** (MST) for the network ABCD may be found using Prim's algorithm or Kruskal's algorithm.

The total length of the MST is 5 + 7 + 14 = 26.

An upper bound for the TSP is 2 × 26 = 52. This corresponds to the route ABCDCBA.

In some cases you might be able to find more than one short-cut that will enable you to reduce the upper bound.

However, an improved (smaller) upper bound can be found by using a **short-cut** from D directly back to A. The route is then ABCDA and the corresponding upper bound is 40.

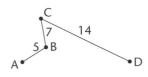

A different approach is to use the **nearest neighbour algorithm**.

Step 1   Choose a starting vertex.

Step 2   Move from your present position to the nearest vertex not yet visited.

Step 3   Repeat step 2 until every vertex has been visited. Return to the start vertex.

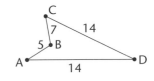

For the network ABCD considered above, the nearest neighbour algorithm gives the same result as using the MST with a short-cut. This won't always be the case.

Finding a lower bound

Step 1   Choose a vertex and delete it along with all of the edges connected to it.

Step 2   Find a minimum spanning tree for the remaining part of the network.

Step 3   Add the length of the MST to the lengths of the two shortest deleted edges.

The value found at step 3 is a lower bound for the travelling salesman problem. It may be possible to find a better (larger) lower bound by choosing to delete a different vertex initially and repeating the process.

Deleting D to start with gives a lower bound of 40. This corresponds to a cycle ABCDA and so represents the solution of the TSP.

Using the network ABCD again and deleting A gives the minimum spanning tree BCD of length 7 + 14 = 21. Adding the two shortest deleted lengths gives a lower bound of 37.

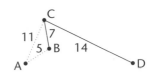

Progress check

1 Find the maximum flow through this network. Verify your answer using the maximum flow-minimum cut theorem.

2 Use a supersource and supersink to find the maximum flow through this network.
Establish the flow through B in this case.

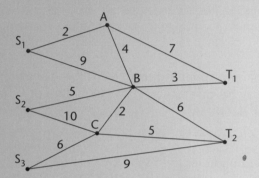

4.6 Dynamic programming

After studying this section you should be able to:

- *understand the concept of dynamic programming*
- *set up a dynamic programming tabulation and use it to solve a problem*

Dynamic programming

Dynamic programming is one of the techniques used to solve problems where some quantity is to be optimised. It involves identifying **stages** and all possible **states** and **actions** within a process. The stages are counted from the end of the process to the beginning.

Working backwards through the stages, decisions are made at each stage that optimise the quantity. Importantly, the calculations carried out must follow from the corresponding optimal value found at the previous stage. In effect, this is applying the **principle of optimality** which states that **any part of an optimal path is itself optimal**.

In the examination, information about the process and the quantity to be optimised may be given in the form of a network or a table.

Example

The network below shows the possible actions and corresponding anticipated profits for a company over the next five months. The figures represent the profit in units of ten thousand pounds, with a negative value indicating a loss due to investment.

(a) Use dynamic programming to maximise the expected profit over the five months.
(b) Give the sequence of vertices in order to achieve the maximum profit.

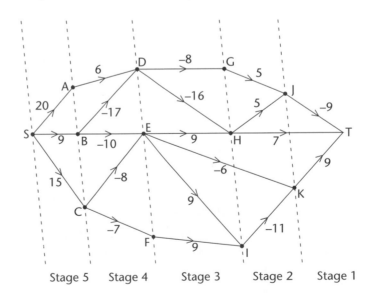

Stage 5   Stage 4   Stage 3   Stage 2   Stage 1

Again, in the examination, an insert is usually provided showing a table to be completed. The table on the next page is typical of the ones used and the coloured entries are typical of the information that would be given on the insert.

(a) An asterisk in the right-hand column is used to indicate the optimal value from the given state.

Stage	State	Action	Value	
1	J	JT	–9	*
	K	KT	9	*
2	G	GJ	5 + –9 = –4	*
	H	HJ	5 + –9 = –4	
		HT	7	*
	I	IK	–11 + 9 = –2	*
3	D	DG	–8 + –4 = –12	
		DH	–16 + 7 = –9	*
	E	EH	9 + 7 = 16	*
		EK	–6 + 9 = 3	
		EI	9 + –2 = 7	
	F	FI	9 + –2 = 7	*
4	A	AD	6 + –9 = –3	*
	B	BD	–17 + –9 = –26	
		BE	–10 + 16 = 6	*
	C	CE	–8 + 16 = 8	*
		CF	–7 + 7 = 0	
5	S	SA	20 + –3 = 17	
		SB	9 + 6 = 15	
		SC	15 + 8 = 23	*

JT and KT are the only possible paths from J and K, so they are automatically optimal.

The value for GJ is calculated as 5 (shown on GJ) plus –9 (the optimal value from J). This is the only value from G so it is automatically optimal.

There are two possible moves from H. The optimal value is 7.

There are three possible moves from E. The value for EH, for example, is 9 (shown on EH) plus 7 (the optimum value for H). This gives 16, which turns out to be the optimal value from E.

F is the last vertex to consider from Stage 3.

The three vertices to consider in Stage 4 are A, B and C.

The highest value found at the final stage is the overall maximum value. In this case this is 23 which represents 23 × £10 000.

The maximum expected profit over the next five months is £230 000.

(b) The sequence of vertices that maximises the profit is SCEHT.

Different types of problem

The dynamic programming technique may be used to solve various types of problem. The way in which the table is used remains the same, but the way in which the optimum value is selected may be different. To find the minimum distance through a network, for example, select the minimum value for each state rather than the maximum as in the example above.

In particular, two types that you need to be aware of are the **maximin** and **minimax**.

The optimal values for the maximin are found by first selecting the minimum value for each action and then choosing the maximum of these to represent the state.

To illustrate how the maximin works, here is part of a network and the corresponding part of a table.

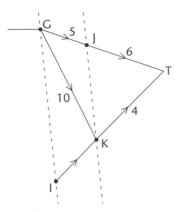

Stage	State	Action	Value	
1	J	JT	6	*
	K	KT	4	*
2	G	GJ	Min(5, 6) = 5	*
		GK	Min(10, 4) = 4	

JT and KT are the only possible paths from J and K, so they are automatically optimal.

The value for GJ is the minimum of 5 (shown on the arc GJ) and 6 (the optimum from J at the previous stage).

The value for GK is the minimum of 10 (shown on the arc GJ) and 4 (the optimum from K at the previous stage).

There are two values for the state G and the maximum of these (5) is shown with an asterisk as this is the optimum for G.

In a similar way, the optimal values for the minimax are found by first selecting the maximum value for each action and then choosing the minimum of these to represent the state.

The table shows the minimax values for the part of the network given above.

Stage	State	Action	Value	
1	J	JT	6	*
	K	KT	4	*
2	G	GJ	Max(5, 6) = 6	*
		GK	Max(10, 4) = 10	

Progress check

1 The diagram shows part of a network. Find the maximin value for the initial state A.

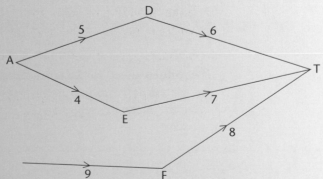

2 Find the minimax value for the initial state A in the diagram in question 1.

2 6
1 5

Sample questions and model answers

1

Ashley and Emma play a two-person zero-sum game in which they each consider three possible strategies. The table shows the pay-off matrix for Ashley.

		Emma		
		X	Y	Z
Ashley	A	5	−3	3
	B	4	1	−2
	C	−3	6	−4

(a) Find the play-safe strategy for each player and show that the game does not have a stable solution.

(b) Set up a linear programming problem to find Ashley's optimal mixed strategy. You are not required to solve it.

(a) The minimum row values are −3, −2 and −4 in Ashley's pay-off matrix. The maximum of these is −2 and so the play-safe strategy for Ashley is strategy B.

The minimum column values in Emma's pay-off matrix are −5, −6 and −3. The maximum of these is −3 and so the play-safe strategy for Emma is strategy Z.

		Emma		
		X	Y	Z
	A	5	−3	3
Ashley	B	4	1	−2
	C	−3	6	−4

		Emma		
		X	Y	Z
	A	−5	3	−3
Ashley	B	−4	−1	2
	C	3	−6	4

The values used to determine the play-safe strategies do not have the corresponding positions in the pay-off matrices so there is no stable solution. Alternatively, $-3 + -2 \neq 0$ and so, again, there is no stable solution.

Sample questions and model answers (continued)

The values need to be made positive by adding a fixed value to each one.

(b) Adding 5 to each of the figures in Ashley's pay-off matrix gives:

10	2	8
9	6	3
2	11	1

Using a, b and c to represent the probability that Ashley uses strategy A, B or C respectively, the linear programming problem is formulated as:

Maximise P subject to $a + b + c \leqslant 1$ where $a \geqslant 0$, $b \geqslant 0$, $c \geqslant 0$ and

$$P - 10a - 9b - 2c \leqslant 0$$
$$P - 2a - 6b - 11c \leqslant 0$$
$$P - 8a - 3b - c \leqslant 0$$

and P represents the value of the game.

Practice examination questions

1   A and B are the players in a two-person zero-sum game. Each may use one of two strategies when a game is played. The pay-off matrix for A is shown below.

Player A	Player B		
		Y	Z
	W	4	8
	X	6	3

(a) Find the best mixed strategy for each player.

(b) Find the value of the game.

2   Richard and Judy play a zero-sum game. The pay-off matrix for Richard is shown in the table.

Richard	Judy			
		X	Y	Z
	A	4	3	−4
	B	−3	6	8
	C	2	1	−6

(a) Find the play-safe strategy for each player.

(b) Show that there is no saddle point for the game.

(c) Explain why strategy C will not be part of Richard's mixed strategy.

(d) Find Richard's optimal mixed strategy.

Practice examination questions *(continued)*

3 A linear programming problem is formulated as:

Maximise $\quad P = x + 2y + 3z$,

subject to $\quad 2x + 6y + z \leqslant 10$

$\qquad\qquad x + 4y + 5z \leqslant 16$

and $\qquad\quad x \geqslant 0,\ y \geqslant 0,\ z \geqslant 0$.

(a) Set up an initial Simplex tableau to represent the problem.

(b) Use the Simplex algorithm to solve the problem.
State the corresponding values of x, y and z.

(c) Explain how you know that the optimum solution has been found.

4 An expedition is planned across a desert from S to T. The network shows the possible routes and distances in km between places where water and supplies are available. The optimal route is the minimax route that minimises the maximum distance between supply points.

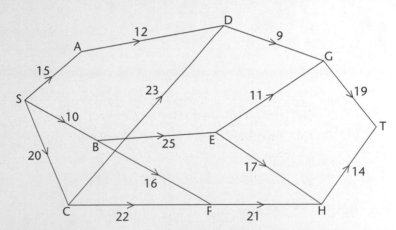

By completing the following table, or otherwise, use dynamic programming, working backwards from T, to find the minimax route and give the maximum distance between supply points on the route.

Stage	State	Action	Value	
1	G	GT	19	*
	H	HT	14	*
2	D	DG	Max(9, 19) = 19	
	E	EH		
	F	FH		
3	A	AD		
	B	BE		
		BF		
4				

Practice examination answers

Core 3 and Core 4 (Pure Mathematics)

1 $\dfrac{1}{(1+x)^2} = (1+x)^{-2}$

$= 1 + (-2)x + \dfrac{(-2)(-3)}{2!}x^2 + \ldots$

$= 1 - 2x + 3x^2 + \ldots \quad |x| < 1$

$\sqrt{4+x} = \sqrt{4\left(1 + \dfrac{x}{4}\right)}$

$= 4^{\frac{1}{2}}\left(1 + \dfrac{x}{4}\right)^{\frac{1}{2}}$

$= 2\left(1 + \dfrac{x}{4}\right)^{\frac{1}{2}}$

$= 2\left(1 + \left(\tfrac{1}{2}\right)\left(\dfrac{x}{4}\right) + \dfrac{\left(\frac{1}{2}\right)\left(-\frac{1}{2}\right)}{2!}\left(\dfrac{x}{4}\right)^2 + \ldots\right)$

$= 2 + \dfrac{x}{4} - \dfrac{x^2}{64} + \ldots \quad |x| < 4$

$\therefore \quad f(x) = 1 - 2x + 3x^2 + 2 + \dfrac{x}{4} - \dfrac{x^2}{64} + \ldots$

$= 3 - 1\tfrac{3}{4}x + 2\tfrac{63}{64}x^2 + \ldots \quad |x| < 1$

Ignoring terms in x^3 and higher, $f(x) \approx a + bx + cx^2$, with $a = 3$, $b = -1\tfrac{3}{4}$, $c = 2\tfrac{63}{64}$.

The restriction is $|x| < 1$.

2 (a) $y = \cos^2 x \sin x$

$\dfrac{dy}{dx} = \cos^2 x(\cos x) + \sin x(-2\sin x \cos x)$

$= \cos^3 x - 2\cos x \sin^2 x$

$= \cos^3 x - 2\cos x(1 - \cos^2 x)$

$= \cos^3 x - 2\cos x + 2\cos^3 x$

$= 3\cos^3 x - 2\cos x$

(b) $u = \cos x \Rightarrow \dfrac{du}{dx} = -\sin x$

$\dfrac{dx}{du} = -\dfrac{1}{\sin x} \Rightarrow \sin x \dfrac{dx}{du} = -1$

$\displaystyle\int \cos^2 x \sin x\, dx$

$= \displaystyle\int \cos^2 x \sin x \dfrac{dx}{du}\, du$

$= \displaystyle\int (-u^2)\, du$

$= -\tfrac{1}{3}u^3 + c$

$= -\tfrac{1}{3}\cos^3 x + c$

(This can be also be done by direct recognition, see page 44.)

3 $\dfrac{dy}{dx} = \dfrac{2x(x+1) - 1(x^2 - 4)}{(x+1)^2}$

$= \dfrac{x^2 + 2x + 4}{(x+1)^2}$

When $x = 2$, $\dfrac{dy}{dx} = \dfrac{4}{3}$

$\therefore$ gradient of normal $= -\tfrac{3}{4}$

Equation of normal at (2, 0):

$y - 0 = -\tfrac{3}{4}(x - 2)$

$4y = -3x + 6$

$3x + 4y - 6 = 0$

$\therefore \quad a = 3,\ b = 4$ and $c = -6$.

4 (a) $g(x) = (x+1)(x-1) = x^2 - 1$.

Since $x^2 \geqslant 0$ the range of g is $x \geqslant -1$.

(b) $fg(x) = f(x^2 - 1)$

$= 5(x^2 - 1) - 7$

$= 5x^2 - 12$.

(c) $fg(x) = f(x)$

$\Rightarrow 5x^2 - 12 = 5x - 7$

$\Rightarrow 5x^2 - 5x - 5 = 0$

$\Rightarrow x^2 - x - 1 = 0$

$\Rightarrow x = \dfrac{1 \pm \sqrt{5}}{2}$.

5 (a) $R\sin(x + a) \equiv 15\sin x + 8\cos x$

$\Rightarrow R\cos a = 15$ and $R\sin a = 8$ (a is acute)

$\Rightarrow R = \sqrt{15^2 + 8^2} = 17$

$\tan a = \tfrac{8}{15} \Rightarrow a = 28.07°$

So $15\sin x + 8\cos x \equiv 17\sin(x + 28.07°)$.

(b) Max value is 17.

(c) $15\sin x + 8\cos x = 12$

$\Rightarrow 17\sin(x + 28.07°) = 12$.

$\sin^{-1}(\tfrac{12}{17}) = 44.90°$, $180° - 44.90° = 135.10°$

$x + 28.07 = 44.90° \Rightarrow x = 16.8°$

$x + 28.07 = 135.10° \Rightarrow x = 107.0°$

Core 3 and Core 4 (Pure Mathematics) (continued)

6 $\quad x\dfrac{dy}{dx} = y + yx \Rightarrow x\dfrac{dy}{dx} = y(1+x)$

Separating the variables:

$\dfrac{1}{y}\dfrac{dy}{dx} = \dfrac{1+x}{x}$

$\dfrac{1}{y}\dfrac{dy}{dx} = \dfrac{1}{x} + 1$

$\displaystyle\int \dfrac{1}{y}\,dy = \int \left(\dfrac{1}{x}+1\right)dx$

$\therefore \quad \ln y = \ln x + x + c$

When $x = 2$, $y = 4$,

$\Rightarrow \quad \ln 4 = \ln 2 + 2 + c$

$\Rightarrow \quad\quad c = \ln 2 - 2 \quad (\ln 4 = 2\ln 2)$

So $\quad \ln y = \ln x + x + \ln 2 - 2$

$\ln\left(\dfrac{y}{2x}\right) = x - 2$

$\dfrac{y}{2x} = e^{x-2}$

$y = 2xe^{x-2}$

7 (a) (i) $\quad y = (4x-3)^9$

$\Rightarrow \dfrac{dy}{dx} = 9(4x-3)^8 \times 4$

$\Rightarrow \dfrac{dy}{dx} = 36(4x-3)^8.$

(ii) $\quad y = \ln(5-2x)$

$\dfrac{dy}{dx} = \dfrac{1}{5-2x} \times (-2)$

$= \dfrac{-2}{5-2x} \quad \left(\text{you can write this as } \dfrac{2}{2x-5}\right).$

(b) (i) $\displaystyle\int (4x-3)\,dx = \dfrac{1}{36}(4x-3)^9 + C.$

(ii) $\displaystyle\int_3^4 \dfrac{1}{5-2x}\,dx = -\dfrac{1}{2}\int_3^4 \dfrac{-2}{5-2x}\,dx$

$= -\dfrac{1}{2}\left[\ln|5-2x|\right]_3^4$

$= -\dfrac{1}{2}\left(\ln|-3| - \ln|-1|\right)$

$= -\dfrac{1}{2}\ln 3.$

8 (a) $\displaystyle\int_0^1 xe^{-2x}\,dx = \left[x(-\tfrac{1}{2}e^{-2x})\right]_0^1 - \int_0^1 (-\tfrac{1}{2}e^{-2x})\,dx$

$= -\tfrac{1}{2}e^{-2} + \tfrac{1}{2}\int_0^1 e^{-2x}\,dx$

$= -\tfrac{1}{2}e^{-2} + \tfrac{1}{2}\left[(-\tfrac{1}{2}e^{-2x})\right]_0^1$

$= -\tfrac{1}{2}e^{-2} + \tfrac{1}{2}(-\tfrac{1}{2}e^{-2} - (-\tfrac{1}{2}))$

$= -\tfrac{3}{4}e^{-2} + \tfrac{1}{4}$

(b) (i) $\quad y = xe^{-x}$

$\dfrac{dy}{dx} = x(-e^{-x}) + e^{-x} = e^{-x}(1-x)$

$\dfrac{dy}{dx} = 0$ when $x = 1$

When $x = 1$, $y = e^{-1} \Rightarrow A$ is point $(1, e^{-1})$.

(ii) $\quad V = \pi\displaystyle\int_0^1 y^2\,dx = \pi\int_0^1 x^2 e^{-2x}\,dx$

$= \pi\left(\left[x^2(-\tfrac{1}{2}e^{-2x})\right]_0^1 - \int_0^1 -\tfrac{1}{2}e^{-2x}(2x)\,dx\right)$

$= \pi\left(-\tfrac{1}{2}e^{-2} + \int_0^1 xe^{-2x}\,dx\right)$

$= \pi(-\tfrac{1}{2}e^{-2} - \tfrac{3}{4}e^{-2} + \tfrac{1}{4})$ (using part (a))

$= \dfrac{\pi}{4}(1 - 5e^{-2})$

9 $\quad \dfrac{dP}{dt} \propto P \Rightarrow \dfrac{dP}{dt} = kP$, where $k > 0$.

Separating the variables

$\displaystyle\int \dfrac{1}{P}\,dP = \int k\,dt$

$\Rightarrow \quad \ln P = kt + c$

$\Rightarrow \quad\quad P = e^{kt+c}$

$\Rightarrow \quad\quad P = Ae^{kt}$ (where $A = e^c$)

$t = 0$, $P = 500 \Rightarrow 500 = A$

$t = 10$, $P = 1000 \Rightarrow 1000 = 500e^{10k}$

$\therefore \quad\quad 2 = e^{10k}$

$\Rightarrow \quad 10k = \ln 2$

$\Rightarrow \quad\quad k = 0.1\ln 2$

$\therefore \quad P = 500e^{(0.1\ln 2)t}$

When $t = 20$,

$P = 500e^{0.1\ln 2 \times 20} = 500 \times 4 = 2000$

So the population is 2000 when $t = 20$.

Core 3 and Core 4 (Pure Mathematics) (continued)

10 (a)
$$x^3 + xy + y^2 = 7$$
$$3x^2 + x\frac{dy}{dx} + y + 2y\frac{dy}{dx} = 0$$

At (1, 2)
$$3 + \frac{dy}{dx} + 2 + 4\frac{dy}{dx} = 0$$
$$\Rightarrow 5\frac{dy}{dx} + 5 = 0 \text{ i.e. } \frac{dy}{dx} = -1$$

At (1, 2) gradient of normal = 1

Equation of normal is
$$y - 2 = 1(x - 1)$$
i.e. $y = x + 1$

(b) $x = t^2$, $y = t + \dfrac{1}{t}$.

$$\frac{dx}{dt} = 2t$$

$$\frac{dy}{dt} = 1 - \frac{1}{t^2} = \frac{t^2 - 1}{t^2}$$

$$\frac{dy}{dx} = \frac{dy}{dt} \times \frac{dt}{dx} = \frac{t^2 - 1}{2t^3}$$

$$\frac{dy}{dx} = 0 \text{ when } t^2 - 1 = 0 \Rightarrow t = \pm 1$$

When $t = 1$, coordinates are (1, 2).

When $t = -1$, coordinates are (1, -2).

The stationary points are at (1, 2) and (1, -2).

11 (a) $\sin 2x = 0$ when $2x = 0$, π, 2π, ...

i.e. when $x = 0$, $\frac{1}{2}\pi$, π, ...

$\therefore a = \frac{1}{2}\pi$

(b) $A = \displaystyle\int_0^{\frac{1}{2}\pi} y\,dx = \int_0^{\frac{1}{2}\pi} \sin 2x\,dx$

$= -\frac{1}{2}\left[\cos 2x\right]_0^{\frac{1}{2}\pi}$

$= -\frac{1}{2}(-1 - 1)$

$= 1$

Area of R is 1 square unit.

(c) $V = \pi \displaystyle\int_0^{\frac{1}{2}\pi} y^2\,dx$

$= \pi \displaystyle\int_0^{\frac{1}{2}\pi} \sin^2 2x\,dx$

$= \frac{1}{2}\pi \displaystyle\int_0^{\frac{1}{2}\pi} (1 - \cos 4x)\,dx$

$= \frac{1}{2}\pi\left[x - \frac{1}{4}\sin 4x\right]_0^{\frac{1}{2}\pi}$

$= \frac{1}{2}\pi(\frac{1}{2}\pi - 0 - 0)$

$= \frac{1}{4}\pi^2$

The volume generated is $\frac{1}{4}\pi^2$ cubic units.

12 (a) Referring to origin (0, 0)

$\mathbf{a} = 2\mathbf{i} + 2\mathbf{j}$ and $\mathbf{b} = 2\mathbf{i} + 3\mathbf{j} + \mathbf{k}$

Also $\mathbf{b} - \mathbf{a} = \mathbf{j} + \mathbf{k}$

A vector equation for AB is

$\mathbf{r} = \mathbf{a} + \mu(\mathbf{b} - \mathbf{a})$ where μ is a scalar parameter

$\therefore \quad \mathbf{r} = 2\mathbf{i} + 2\mathbf{j} + \mu(\mathbf{j} + \mathbf{k})$

(b) If lines intersect, there are unique values for μ and λ for which

$2\mathbf{i} + 2\mathbf{j} + \mu(\mathbf{j} + \mathbf{k}) = 2\mathbf{j} + \mathbf{k} + \lambda(\mathbf{i} + \mathbf{k})$

Equating coefficients of $\mathbf{i}$, $\mathbf{j}$ and $\mathbf{k}$, in turn gives:

$2 = \lambda$

$2 + \mu = 2$

$\mu = 1 + \lambda$

There are no values of μ and λ that satisfy all three equations, so the lines have no common point.

(c) The angle between the lines is the angle between their direction vectors.

$|\mathbf{j} + \mathbf{k}| = \sqrt{2}$ and $|\mathbf{i} + \mathbf{k}| = \sqrt{2}$

$(\mathbf{j} + \mathbf{k}).(\mathbf{i} + \mathbf{k}) = 1$

Let the angle be θ, then

$$\cos\theta = \frac{1}{\sqrt{2}\sqrt{2}} = 0.5$$

$\Rightarrow \quad \theta = 60°$

Statistics 2

1 (a) If X is number of people with the health problem, then $X \sim B(80, 0.01)$. n is large and p is small, so $X \sim Po(np)$ i.e. $X \sim Po(0.8)$ approximately.

$$P(X < 3) = e^{-0.8} + 0.8\,e^{-0.8} + \frac{0.8^2}{2!}\,e^{-0.8}$$

$$= 0.953 \text{ (3 s.f.)}$$

(b) If X is number of people with the health problem in a sample of size n, $X \sim B(n, 0.01)$.
If $n > 50$, $X \sim P(0.01n)$ approximately.

$$P(X \geqslant 1) = 1 - P(X = 0) = 1 - e^{-0.01n}$$

So $P(X \geqslant 1) > 0.95 \Rightarrow 1 - e^{-0.01n} > 0.95$
i.e. $e^{-0.01n} < 0.05$

By trial and improvement, $e^{-2.99} = 0.0502 > 0.05$, $e^{-3} = 0.049 < 0.05$, so $0.01n = 3 \Rightarrow n = 300$. (Log theory could be used here.) The minimum number in the sample should be 300.

2 (a) $f(x) = k$, $2 \leqslant x \leqslant 12$
Total probability = 1
$\Rightarrow 10 \times k = 1$, i.e. $k = \frac{1}{10}$
$\therefore \; f(x) = \frac{1}{10}$, $2 \leqslant x \leqslant 12$

(b) By symmetry, $E(X) = \frac{1}{2}(2 + 12) = 7$

$$E(X^2) = \frac{1}{10}\int_2^{12} x^2 \, dx = \frac{1}{10}\left[\frac{x^3}{3}\right]_2^{12} = 57\tfrac{1}{3}$$

$$Var(X) = 57\tfrac{1}{3} - 7^2 = 8\tfrac{1}{3}$$

or $Var(X) = \frac{1}{12}(12 - 2)^2 = 8.333\ldots$ (formula)

Standard deviation $= \sqrt{8\tfrac{1}{3}} = 2.887$ (3 d.p.)

$P(7 - 2.887 < X < 7 + 2.887)$
$= P(4.113 < X < 9.887)$
$= 0.1 \times (9.887 - 4.113)$
$= 0.58$ (2 d.p.)

3 Let X be the number of trees with the infestation. Assuming that trees are infested independently, with probability p, $X \sim B(n, p)$

$H_0: p = 0.35$
$H_1: p > 0.35$

If $p = 0.35$, $X \sim B(10, 0.35)$
At the 10% level,
H_0 is rejected if $P(X \geqslant 6) < 0.1$.

Using cumulative binomial tables,
$P(X \geqslant 6) = 1 - P(X \leqslant 5)$
$= 1 - 0.9051 = 0.0949$

Since $P(X \geqslant 6) < 0.1$, H_0 is rejected.

There is evidence, at the 10% level, that the percentage of trees that are infested is greater than 35%, indicating that the trees should be felled.

Statistics 2 (continued)

4 (a) $\displaystyle\int_{\text{all }x} f(x)\,dx = 1$

    so $\displaystyle 1 = k\int_0^1 (x - x^3)\,dx$

$$= k\left[\frac{x^2}{2} - \frac{x^4}{4}\right]_0^1$$

$$= k\left(\frac{1}{2} - \frac{1}{4}\right)$$

$$= \frac{1}{4}k$$

    $\therefore$ $k = 4$

(b) $\displaystyle \mu = E(X) = \int_0^1 xf(x)\,dx = 4\int_0^1 x(x - x^3)\,dx$

$$= 4\int_0^1 (x^2 - x^4)\,dx$$

$$= 4\left[\frac{x^3}{3} - \frac{x^5}{5}\right]_0^1$$

$$= 4\left(\frac{1}{3} - \frac{1}{5}\right)$$

$$= \frac{8}{15}$$

(c) $\displaystyle P(X < 0.5) = 4\int_0^{0.5} (x - x^3)\,dx$

$$= 4\left[\frac{x^2}{2} - \frac{x^4}{4}\right]_0^{0.5}$$

$$= 0.4375$$

(d) $\displaystyle E(X^2) = \int_0^1 x^2 f(x)\,dx = 4\int_0^1 x^2(x - x^3)\,dx$

$$= 4\int_0^1 (x^3 - x^5)\,dx$$

$$= 4\left[\frac{x^4}{4} - \frac{x^6}{6}\right]_0^1$$

$$= \frac{1}{3}$$

    $\text{Var}(X) = E(X^2) - \mu^2$

$$= \frac{1}{3} - \left(\frac{8}{15}\right)^2$$

$$= 0.0489 \ (3 \text{ s.f.})$$

5 (a) $F(2) = 1$

    $\Rightarrow c(8 - 4 - 1) = 1$ so $c = \frac{1}{3}$

(b) $P(X > 1\frac{1}{2}) = 1 - F(1\frac{1}{2})$

$$= 1 - \tfrac{1}{3}(4 \times 1\tfrac{1}{2} - (1\tfrac{1}{2})^2 - 1)$$

$$= \tfrac{1}{12}$$

(c) For $0 \leqslant x \leqslant 1$,

$$f(x) = \frac{d}{dx}\left(\tfrac{2}{3}x\right) = \tfrac{2}{3}$$

    For $1 \leqslant x \leqslant 2$,

$$f(x) = \frac{d}{dx}\left(\tfrac{1}{3}(4x - x^2 - 1)\right)$$

$$= \tfrac{1}{3}(4 - 2x) = \tfrac{2}{3}(2 - x)$$

    For $x \leqslant 0$, $x \geqslant 2$, $f(x) = 0$

(d) $\displaystyle E(X) = \int_0^1 \frac{2}{3}x\,dx + \int_1^2 \frac{2}{3}(2x - x^2)\,dx$

$$= \left[\frac{1}{3}x^2\right]_0^1 + \frac{2}{3}\left[x^2 - \frac{x^3}{3}\right]_1^2$$

$$= \tfrac{1}{3} + \tfrac{2}{3}(4 - \tfrac{8}{3} - (1 - \tfrac{1}{3}))$$

$$= \tfrac{7}{9}$$

Mechanics 2

1 (a) Using $y = x\tan\theta - \dfrac{gx^2}{2v^2}(1 + \tan^2\theta)$

gives $2.5 = 20\tan\theta - \dfrac{9.8 \times 20^2}{2 \times 25^2}(1 + \tan^2\theta)$

$3.136\tan^2\theta - 20\tan\theta + 5.636 = 0$

$\tan\theta = 6.082\ldots$ or $\tan\theta = 0.2954\ldots$

$\tan^{-1}(0.2954\ldots) = 16.5^0$

so a possible angle of projection is 16.5^0.

(b) Vertically: using $s = ut + \frac{1}{2}at^2$ gives

$-0.5 = 25\sin 16.5^0\,t - 4.9t^2$

$4.9t^2 - 25\sin 16.5^0\,t - 0.5 = 0$

$t = 1.516\ldots$ or $t = -0.067\ldots$

since $t > 0$

$d = 25\cos 16.5^0 \times 1.516 - 20$

The distance is 16.3 m to 3 s.f.

2 (a) $x = t^3 - 6t^2 + 3$
$\dot{x} = 3t^2 - 12t = 3t(t - 4)$
When $t = 0$, $\dot{x} = 0$ so initially at rest. At rest again
when $t = 4$
$4^3 - 6 \times 4^2 + 3 = -29$
Max distance in negative direction is 29 m.

(b) The particle is at its initial position when
$t^3 - 6t^2 = 0 \Rightarrow t^2(t - 6) = 0$
Particle passes through initial position when $t = 6$.
This gives $\dot{x} = 3 \times 6^2 - 12 \times 6 = 36\ \text{ms}^{-1}$.

3 (a) $\mathbf{r} = (4t^2 + 5)\mathbf{i} + 2t^3\mathbf{j}$
$\dot{\mathbf{r}} = 8t\mathbf{i} + 6t^2\mathbf{j}$

When $t = 1$, $\dot{\mathbf{r}} = 8\mathbf{i} + 6\mathbf{j}$ so the speed $= 10\ \text{ms}^{-1}$.

(b) P is moving parallel to
$\mathbf{i} + 3\mathbf{j} \Rightarrow 6t^2 = 24t \Rightarrow 6t(t - 4) = 0$
Since $t > 0$ this gives $t = 4$.

4 (a) 3 cm (by symmetry).

(b) $\bar{y} = \dfrac{24 \times 2 - 4 \times 3}{20} = 1.8$ cm

(c)

$\theta = \tan^{-1}\left(\frac{3}{1.8}\right) = 59.0^0$

5 (a) Area rectangle : area triangle = 3 : 1

so mass of rectangle = 0.6 kg
mass of triangle = 0.2 kg

$\bar{x} = \dfrac{0.6 \times 7.5 + 0.2 \times 18\frac{1}{3}}{0.8} = 10.2$ cm

(b)

$15R_B = 10.2 \times 8 \Rightarrow R_B = 5.44$ N
$R_A = 8 - 5.44 = 2.56$ N

Mechanics 2 (continued)

6 (a)

Conservation of momentum:

$0.9 \times 4 - 0.3 \times 2 = 0.9u + 0.3v$

so $3u + v = 10$ (1)

Newton's law of restitution:

$v - u = \frac{1}{3} \times 6 = 2$ (2)

From (1) and (2) $4u = 8$
giving $u = 2$, $v = 4$.
The speeds of A and B after the collision are 2 ms^{-1} and 4 ms^{-1} respectively.

(b) Impulse = change in momentum

$= 0.3 \times 4 - 0.3 \times (-2)$

$= 1.8$ Ns

(c) Total KE before impact

$= \frac{1}{2} \times 0.9 \times 4^2 + \frac{1}{2} \times 0.3 \times 2^2 = 7.8$ J

Total KE after impact

$= \frac{1}{2} \times 0.9 \times 2^2 + \frac{1}{2} \times 0.3 \times 4^2 = 4.2$ J

Loss of mechanical energy = 3.6 J

7

Conservation of momentum:

$3mu - 2mu = 3mv + mw$

giving: $3v + w = u$ (1)

Newton's law of restitution:

     $w - v = 3ue$ (2)

(1)–(2) gives $4v = u(1 - 3e)$

so $v = \frac{u}{4}(1 - 3e)$

$v < 0 \Rightarrow (1 - 3e) < 0$

$\Rightarrow e > \frac{1}{3}$

8 (a) The tension is perpendicular to the direction of motion.

(b) The acceleration is not constant.

(c) Maximum speed occurs at lowest point.

Loss of PE = gain in KE

$mgl = \frac{1}{2}mv^2$

so, maximum speed is given by $v = \sqrt{2gl}$

9 (a)

Resolving parallel to the plane:

$F = 900 \times 9.8 \times \frac{1}{15} + 800$

$= 1388$

The driving force is 1388 N

(b) $P = Fv = 1388 \times 20 = 27760$

Power $= 27.76$ kW

(c) Using $F = ma$:

gives: $1388 - 800 = 900a \Rightarrow a = 0.65\ ms^{-2}$.

Mechanics 2 *(continued)*

10   Momentum of ball before impact
= $0.2 \times 20 = 4$ Ns

Momentum of ball after impact

= $0.2 \times 5 = 1$ Ns

The magnitude of the impulse is given by:

$$I^2 = 1^2 + 4^2 = 17 \Rightarrow I = \sqrt{17}$$

$\tan \theta = \frac{1}{4} \Rightarrow \theta = 14.0^0$

The impulse has magnitude $\sqrt{17}$ Ns at an angle of 14.0^0 above the horizontal.

11   (a)   The velocity at time t is given by:
$\dot{x} = 12t^2 - 6t = 6t(2t - 1)$

    (b) $\dot{x} = 0$ when $t = 0$ or $t = 0.5$ so the particle changes direction when $t = 0.5$

        $t = 0 \Rightarrow x = 7$

        $t = 0.5 \Rightarrow x = 6.75$

        The distance travelled is 0.25 m.

    (c)   P is at its starting point when

        $4t^3 - 3t^2 = 0$

      $\Rightarrow t^2(4t - 3) = 0$

      $\Rightarrow t = 0$, or $t = 0.75$

        P returns to its start point after 0.75 s

Decision Mathematics 2

1 (a) Assume A uses Strategy W with probability p and strategy X with probability $(1 - p)$.

Expected gain for A is:

$4p + 6(1 - p)$ if B uses strategy Y.

$8p + 3(1 - p)$ if B uses strategy Z.

Optimal when

$4p + 6(1 - p) = 8p + 3(1 - p)$

$\Rightarrow -2p + 6 = 5p + 3$

$\Rightarrow p = \frac{3}{7}$

Assume B uses strategy Y with probability q and strategy Z with probability $(1 - q)$.

Expected loss for B is:

$4q + 8(1 - q)$ if A uses strategy W.

$6q + 3(1 - q)$ if A uses strategy X.

Optimal when

$4q + 8(1 - q) = 6q + 3(1 - q)$

$\Rightarrow -4q + 8 = 3q + 3$

$\Rightarrow q = \frac{5}{7}$

The optimal strategy for A is to use strategy W with probability $\frac{3}{7}$ and strategy X with probability $\frac{4}{7}$.

The optimal strategy for B is to use strategy Y with probability $\frac{5}{7}$ and strategy Z with probability $\frac{2}{7}$.

(b) The value of the game is $5\frac{1}{7}$.

2 (a) The minimum row values are −4, −3 and −6. The maximum of these is −3 showing that the play-safe strategy for Richard is strategy B.

The minimum of the negatives of the column values are −4, −6 and −8. The maximum of these is −4 showing that the play-safe strategy for Judy is strategy X.

(b) $-3 + (-4) \neq 0$ so there is no saddle point.

(c) Comparing the row values shows that Richard's pay-offs are *always* less with strategy C than with strategy A. It follows that strategy C will not be a part of his mixed strategy.

(d) Assume Richard uses strategy A with probability p and strategy B with probability $(1 - p)$.

The expected gains for Richard based on the strategy used by Judy are:

X: $4p - 3(1 - p) = 7p - 3$

Y: $3p + 6(1 - p) = -3p + 6$

Z: $-4p + 8(1 - p) = -12p + 8$

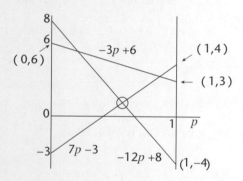

Optimal when $7p - 3 = -12p + 8$

$\Rightarrow p = \frac{11}{19}$

Richard's optimal strategy is to use strategy A with probability $\frac{11}{19}$ and strategy B with probability $\frac{8}{19}$.

Decision Mathematics 2 (continued)

3 (a) The initial tableau is:

P	x	y	z	s	t	val
1	−1	−2	−3	0	0	0
0	2	6	1	1	0	10
0	1	4	5	0	1	16

(b) The first iteration produces:

P	x	y	z	s	t	val
1	$-\frac{2}{5}$	$\frac{2}{5}$	0	0	$\frac{3}{5}$	$9\frac{3}{5}$
0	$\frac{9}{5}$	$\frac{26}{5}$	0	1	$-\frac{1}{5}$	$6\frac{4}{5}$
0	$\frac{1}{5}$	$\frac{4}{5}$	1	0	$\frac{1}{5}$	$3\frac{1}{5}$

The second iteration produces:

P	x	y	z	s	t	val
1	0	$\frac{14}{9}$	0	$\frac{2}{9}$	$\frac{5}{9}$	$11\frac{1}{9}$
0	1	$\frac{26}{9}$	0	$\frac{5}{9}$	$-\frac{1}{9}$	$3\frac{7}{9}$
0	0	$\frac{2}{9}$	1	$-\frac{1}{9}$	$\frac{2}{9}$	$2\frac{4}{9}$

The maximum value of P is $11\frac{1}{9}$.
This occurs when $x = 3\frac{7}{9}$, $y = 0$, $z = 2\frac{4}{9}$.

(c) This represents the optimum solution because there are no negative values in the objective row of the tableau.

4

Stage	State	Action	Value	
1	G	GT	19	*
	H	HT	14	*
2	D	DG	Max(9, 19) = 19	
	E	EH	Max(17, 14) = 17	*
	F	FH	Max(21,14) = 21	
3	A	AD	Max(12, 19) = 19	*
	B	BE	Max(25, 17) = 25	
		BF	Max(16, 21) = 21	*
	C	CD	Max(23, 19) = 23	
		CF	Max(22, 21) = 22	*
4	S	SA	Max(15, 19) = 19	*
		SB	Max(10, 21) = 21	
		SC	Max(20, 22) = 22	

The optimal minimax route is SADGT.
The maximum distance between supply points on the route is 19 km.

Notes

Index